Frank Julian Sprague

Frank Julian Sprague

ELECTRICAL INVENTOR & ENGINEER

WILLIAM D. MIDDLETON & WILLIAM D. MIDDLETON III

Foreword by JOHN L. SPRAGUE

Indiana University Press ⊚ Bloomington & Indianapolis

This book is a publication of

Indiana University Press
Office of Scholarly Publishing
Herman B Wells Library 350
1320 East 10th Street
Bloomington, Indiana 47405 USA

www.iupress.indiana.edu

This book is printed on acid-free paper.

Manufactured in the United States of America

Library of Congress Cataloging-in-Publication Data

Middleton, William D., date
 Frank Julian Sprague : electrical inventor and engineer / William D. Middleton and William D. Middleton III ; foreword by John L. Sprague.
   p. cm. — (Railroads past and present)
 Includes bibliographical references and index.
 ISBN 978-0-253-35383-2 (cloth : alk. paper)
 1. Sprague, Frank J. (Frank Julian), 1857–1934. 2. Inventors—United States—Biography. 3. Electrical engineers—United States—Biography. 4. Electric railroads—United States—History—20th century. I. Middleton, William D., date II. Title.
 TA140.S7M53 2009
 621.3092—dc22
 [B]
                        2009017091

 2   3   4   5   19   18   17   16   15

FRONTIS Frank Julian Sprague (1857–1934) photographed circa mid-1890s. *Middleton Collection.*
RIGHT *Photo courtesy of Mary Ann Smerk.*

To Professor George M. Smerk, our long-time friend and colleague, whose career in education and transportation have deeply enriched the fields at many levels. His assistance on this and many other projects is much appreciated.

RAILROADS PAST & PRESENT

George M. Smerk, editor

A list of books in the series appears at the end of this volume.

CONTENTS

FOREWORD

A 1932 photograph shows a trim elderly man holding a chubby two-year-old child. The man is well dressed and has a slightly quizzical expression as he regards his armful. His face is narrow and seems constructed of sharp angles and lines. He has a full head of hair, a prominent nose, and a full but well-trimmed mustache. But it is his eyes that grip you. Even in the slightly faded image, behind his metal-rimmed glasses they seem to glitter with intelligence. On the other hand, the child is clearly oblivious to the fact that the arms holding him belong to the man who, at the time, was renowned as "the Father of Electric Traction." This photograph is the only recorded proof I have that my grandfather and I ever met. He died only two years later and unfortunately I have no recollection of either the event or of him.

This is not true of my grandmother, his second wife, Harriet, who outlived her husband by some 35 years. I met and talked with her often as I was growing up in Williamstown, Massachusetts, and she was a warm and loving companion. However, this sweet little old lady was full of surprises. She was a renowned Walt Whitman collector, and also showed a steely ferocity when defending the legacy of her beloved Frank. Her 1947 24-page monograph, *Frank J. Sprague and the Edison Myth,* was, I believe, the first serious attack on the legacy of "the Wizard of Menlo Park."[1] In the book that follows, the Middletons spend a full chapter on the initially cordial and then increasingly contentious relationship between these two men.

As I grew up in Williamstown, Frank Sprague was all around me. There were portraits and photographs and memorabilia of all kinds. One of the favorites that I still have is a plastic molded length of approximately one-inch cable that was part of the first three-wire underground distribution system in the world, and which was installed in Brockton, Massachusetts, in October 1883 by Thomas A. Edison with Frank J. Sprague as the "Resident Engineer in Charge of Construction." Another was a large-scale model of one of his early street railway wheelbarrow motor suspension trucks that resided in my father's office throughout his business career. All of our family members were, and still are, in awe of this famous inventor. Yet as I grew up we almost never talked about him. Recently discovered family letters indicate that, while he was a loving father, Frank Sprague was often too busy or away to spend much time with his children. At the time I also sensed that my father was much closer to his mother and that his feelings about his father were ambivalent. He certainly had a fierce pride in Frank Sprague's obvious genius as an inventor and entrepreneur. Yet he also seemed frustrated by the fact that, when Frank Sprague died in 1934, the modest fortune that he had built over the years, primarily as the result of selling The Sprague Electric Company to General Electric in 1902, had vanished. As far as I know, my father completely supported his mother during the last 35 years of her life. She died in 1969 at the age of 93.

In the late 1950s, Harriet, my father, and his younger brother, Julian, contracted with science writer E. S. Lincoln to write a biography of Frank J. Sprague. Completed in late 1959, it was circulated to more than a dozen different publishers. While reception was lukewarm, it reached all the way to galley proofs with The Bobbs-Merrill Company before discussions fell apart over the publisher's demand for major underwriting by the family. Unhappy with the depiction of Frank Sprague's personal life in the Lincoln biography, they then turned to Frank Rowsome, Jr., for a rewrite. At the time, Rowsome was managing editor of *Popular Science* and had written *Trolley Car Treasury*, a picture book published in 1956.[2] Unfortunately, after circulation to a number of different publishers, the much more readable Rowsome manuscript met the same fate when negotiations with McGraw-Hill collapsed after the publisher made similar underwriting demands. Ironically, although I joined the Research Labs of the second Sprague Electric Company (founded in 1925 by my father as The Sprague Specialties Company) during this same period, I only learned of the existence of these two manuscripts in 1991 while reviewing my father's papers after his death.

In the early 1980s a series of fortunate events occurred that created a renewed and growing interest in my grandfather. First, I found a copy of Harold

C. Passer's *Frank Julian Sprague: Father of Electric Traction*.[3] Almost simultaneously I received a copy of a 1984 French article concerning "La Fin des Sprague," the retirement of the Paris metro line Le Sprague. Having run since 1903, the line was an amalgamation of the Sprague and Thomson railway systems. Even until quite recently, some of its cars were run on holidays, used in nighttime maintenance runs, and were part of an exclusive all-night tour under Paris sponsored by ADEMAS, the Association d'Exploitation du Matériel Sprague. After several trips to France and correspondence with a whole group of new friends both in France and in the United Kingdom, I now recognize that in both of those countries Frank J. Sprague is better known than he is in the United States. I did find an important exception to this observation when, on September 30, 2006, I participated in the celebration "A Century of Third Rail Power" in New York City's Grand Central Station. The third rail electrification system was co-invented by William J. Wilgus and Frank J. Sprague and there I met yet another group of people who know of and revere my grandfather.

But the event that really fueled my now rapidly growing interest was receipt from my father of the six-volume letterbooks given to my grandfather at the 1932 celebration of his 75th birthday. By now I have read them all numerous times and they are frequently referenced by the Middletons in the biography that follows. But at the time, it was the names of those who wrote my grandfather that struck me: President Herbert Hoover, New York governor Franklin D. Roosevelt, movie magnate William C. deMille, authors Oliver Herford and Booth Tarkington, friend and fellow inventor Guglielmo Marconi, General John J. Pershing, Nobel Prize recipient R. A. Millikan, German industrialist Carl von Siemens, fellow inventor Nikola Tesla, and one-time antagonist, General Electric honorary chairman Edwin Wilbur Rice, to name just a few. More than 500 individuals from around the world wrote personal letters with accompanying photographs to Frank J. Sprague: family, friends, associates, competitors, and men and women of arts and letters who became his close friends after he married Harriet. This collection offers an extraordinary celebration of how the world perceived Frank J. Sprague toward the end of his life.

As my grandfather's image grew to heroic proportions, a nagging question kept returning. Why is there still no published biography of Frank J. Sprague? Perhaps he is just too out-of-date or the subject uninteresting. More likely it is because no major corporation based on his work and bearing his name still exists today, nor in the United States is there any electrification system with his name such as Paris's Le Sprague. In the late 1980s my father stubbornly tried a different route and several times attempted to have his father elected to the National Inventors Hall of Fame, again to no avail. Each time the response was: "He did not receive a sufficient number of votes."

Early in the new millennium I began to visit the New York Public Library and to read the Frank J. Sprague Papers, especially his patents. As covered in detail by the Middletons, they are extraordinary and unlike those of any of his competitors. It isn't just the descriptive depth or breadth of what he covered. Often he also included detailed theoretical discussions of just why his inventions worked the way they did, as in the development of the "Sprague Laws" in his early motor patents. And his basic Multiple Unit Control (MU) patent (US #660,065, filed April 30, 1898, and issued October 16, 1900) is a masterpiece. This massive document covers every conceivable aspect of the system and is the reason that General Electric finally bought his third company, The Sprague Electric Company, in 1902.

So it is a very great honor to write a foreword to this long awaited and much needed biography of Frank J. Sprague. It is meticulously researched and, in glorious detail, the Middletons have skillfully portrayed the life and times of this extraordinary man, not just as an engineer, inventor, entrepreneur, and businessman, but also as a man. For he was also a husband, father, and friend to many, and an individual who created great and lasting loyalty in almost everyone who was ever close to him or had the privilege to work with him.

John L. Sprague

ACKNOWLEDGMENTS

At the time of his death in 1934 Frank Sprague was a widely known and celebrated electrical engineer and inventor, and the *New York Herald Tribune* had ranked Sprague with Thomas Edison and Alexander Bell as a remarkable trio of inventors. "Perhaps no three men in human history," said the *Herald Tribune*, "have done more to change the daily lives of human kind." In the years since their death both Edison and Bell have retained the general public's recognition and appreciation of their work. But, in the way that public notice seems to arbitrarily raise one up to great recognition while another seems to disappear from notice, the work and accomplishments of Frank Sprague are now largely forgotten.

For the two writers of this biography, however, the story of Frank Sprague has long been viewed with great appreciation and recognition. For the elder, Sprague's work in electric traction has made him a hero for well over a half century in Middleton Sr.'s lifelong interest in electric railways, while the younger has followed Sprague's phenomenal lifetime of work and accomplishment as an inventor with no less admiration.

The information sources about Frank Sprague are many and varied, but by far the most comprehensive are the Frank J. Sprague Papers held by the Rare Books and Manuscripts Division of the New York Public Library. Collected by his widow, Harriet Sprague, the collection includes some 159 boxes or volumes, dozens of notebooks and memoranda books, and more than 200 volumes of a variety of published material covering almost every aspect of his professional and personal life. Our long and fruitful searches through this material were greatly aided by New York Public Library staff members Thomas Lannon, Susan Waide, Laura Ruttum, Megan O'Shea, and Nasima Hasnat, who without fail located the many dozens of boxes, volumes, and other material that we sought during our many days with the library.

Significant grants of Sprague material were also given to the Engineering Societies Library in New York—later transferred to the Linda Hall Library in Kansas City, Missouri, after the disestablishment of the Engineering Societies Library—and to the Shore Line Trolley Museum in East Haven, Connecticut. The six bound volumes of letters and photographs received from more than 500 friends, family, and associates sent to Frank Sprague on his 75th birthday are held by the rare books library of The Chapin Library at Williams College, Williamstown, Massachusetts.

The immense resources of the University of Virginia Library in Charlottesville, Virginia, have been a continuing resource, particularly for the strong technical publications held by the library's Ivy Stacks, where Ray O'Donohue and Steve Bartlett have never failed to find and bring forth many dozens of volumes of nineteenth- and early twentieth-century engineering publications. In those rare cases when the university library didn't hold something, their interlibrary loan service was almost always able to find it elsewhere.

The library of the Electric Railroaders' Association at Grand Central Terminal in New York City, with much assistance from William K. Guild, was of great help with material concerning electric traction. The Shore Line Trolley Museum offered valuable service on several scores. Librarian Michael Schreiber made available the museum's extensive Sprague collection, while Fred Sherwood demonstrated the rare 1884 Sprague electric motor that he has restored to operation as a major part of the museum's Frank Sprague exhibit. The museum's training director, Jeff Hakner, has been extremely helpful in arranging our visits to the museum, and very carefully reviewed our chapters on electric railway development from the vantage point of an electrical engineer.

In Washington, D.C., the U.S. Patent and Trademark Office provided its extensive records of Frank Sprague's patents. The Library of Congress made available several rare volumes for review, including Frank J. Sprague's 1883 book on the 1882 Crystal Palace electrical exhibition in London, and made available several copies of Sprague's splendid drawings.

The U.S. Naval Academy, with much help from Jennifer M. Erickson, media relations specialist, and Beverly Lyall, archives technician at the Nimitz Library, provided extensive material and photographs concerning life in Annapolis during Sprague's 1874–1878 years as an Academy undergraduate, while the U.S. Naval Academy Alumni Association provided valuable information about Sprague's classmates. The Naval Historical Center at the Washington Navy Yard provided much help with both historical material and photographs.

Insights into Frank Sprague's life in the Naval Academy and through his naval service were provided from an extensive file of letters from Sprague to Miss Mattie Munroe, a young lady in Massachusetts, from 1876 to 1881 that are now held by the Special Collections Department, J. Y. Joyner Library, East Carolina University, in Greenville, North Carolina. We are indebted to Dale Souter of the Special Collections Department for making copies of the letters available, and to Andrea Klarman, our daughter-in-law/sister-in-law, for managing the task of reading the faded old letters for us.

The City of Milford (Connecticut) provided information on the birth and deaths of members of the Sprague family and information about Milford in the mid-nineteenth century. Edward M. Kirby, president of the Sharon Historical Society, provided valuable information on the history of The Maples, summer home of the Spragues for almost 20 years. He also arranged a tour of the splendid building with his daughter, Maureen Doer, who with her husband, Thomas Patrick Dore, Jr., are now the present owners of The Maples.

Extensive historical material about North Adams, Massachusetts, was provided by the North Adams Public Library, with much help being provided by special collections librarian Katharine C. Westwood. Gene S. Carlson, treasurer of North Adams Historical Society provided many historical publications and some splendid photographs of the nineteenth-century city.

Professor Lee E. Gray, an architectural historian and clearly an elevator aficionado, has recently completed a history of the passenger elevator in the nineteenth century, *From Ascending Rooms to Express Elevators,* which told us much about the 1890–1900 period when Frank Sprague was a leader in turning from steam and hydraulic elevators to electric power. Lee's more recent work for the magazine *Elevator World* offers valuable information about Sprague's later efforts to develop the dual elevator system in 1931.

A much appreciated colleague on the book has been Professor George M. Smerk, recently retired from a long career as the head of Indiana University's Institute for Urban Transportation, who knows just about everything associated with electric traction, and who provided very valuable advice after reviewing our manuscript. Brian J. Cudahy, now retired from a career at the U.S. Department of Transportation, provided a number of valuable materials concerning Frank Sprague's history in the U.S. Navy from the U.S. National Archives and his attendance at the Drury High School in North Adams, Massachusetts.

Of special significance to us have been the advice and assistance of John L. Sprague, the son of Robert C. Sprague and a grandson of Frank Sprague. He made available to us extensive quotes from the 1932 Birthday Books, a wide variety of notes, family letters, recollections, family photographs, provided a splendid foreword for the book, and—perhaps most important of all—his enthusiastic support of our work, which has provided us with much insight into the Sprague family. Darnall Burks, a cousin and the Sprague family genealogist, has provided advice concerning Sprague family relationships.

The very capable members of the Indiana University Press staff have all contributed in putting this book in such handsome form. For this, we thank Linda Oblack, Peter Froehlich, Miki Bird, Chandra Mevis, June Silay, Jamison Cockerham, and Tony Brewer.

William D. Middleton
William D. Middleton III

Frank Julian Sprague

Phoenix Bridge Dam, No. Adams, Mass

The big brick and stone mill buildings and the water power from the Hoosac River were the backbones of busy North Adams late in the 1800s. The Phoenix Bridge Dam was located on the South Branch of the Hoosac River just above Main Street on the west end of the downtown area. The dam supplied water power to the Phoenix Mills.   *North Adams Historical Society.*

1

A BOYHOOD IN NEW ENGLAND

Milford, Connecticut, is now a city of more than 50,000 residents, lying some 10 miles southwest of New Haven and stretching along the shores of Long Island Sound. Milford grew large only in the recent past with the growth that followed World War II, but it has been there a long time. What became Milford, named after the English city, was purchased by English settlers from the chief of the local Paugusset tribe in 1639, making it the sixth oldest community in Connecticut. Even today Milford retains much of the character that dates to the nineteenth century and before. The Wepawaug River winds down through the town and into the oyster-rich estuary of Long Island Sound. Just west of the river, Milford's carefully maintained "town green"—the second longest in all New England, boast the residents—stretches a block wide and six blocks long. The green of the square is intermingled with trees and monuments from Milford's—and America's—past.

A century and a half ago Milford had scarcely 2,500 residents, and the working population was occupied with farming, oystering, shipbuilding, and a few industrial plants, while the Long Island Sound shore served as a beach resort for residents of New Haven and Bridgeport. The young David Cummings Sprague came to Milford about 1852 to become a plant superintendent for a hat manufacturing firm, one of many in the southwest Connecticut area centered on Danbury that made the state a major supplier of hats. Born in Wardsboro, Vermont, in 1833, D. C. Sprague was one of ten children born to Joshua Sprague, who was of the eighth American generation descended from

Ralph Sprague. The latter had left England from the hamlet of Upwey in Devonshire in 1628.[1]

Only 19 years old, David Sprague married Frances Julia King in 1842, and the young couple established a home in Milford that was reputed to have been the refuge of two English regicides who fled to New England and were hidden in a Milford cellar in 1661 after being condemned to death for the execution of Charles I.[2] The Sprague family had encountered the profound disappointment of the death of their first child, Sieber or Seaver, who died at birth in April 1856. But three years later the family's first surviving son, Frank Julian Sprague, was born in Milford on July 25, 1857, followed by another son, Charles May Sprague, on April 30, 1860.

The Sprague family was abruptly changed early in 1866, when Frances died suddenly from a hemorrhage on January 31. David Sprague soon decided that he would seek his fortune in the West, while the two boys were taken off to North Adams, Massachusetts, where they would be left in the care of David's older sister, Elvira Betsy Ann Sprague. The loneliness and uncertainty of this abrupt change in their lives was suggested many years later in a reminiscence of her early school life by Mrs. Susan Amelia Shove, who wrote in the *Milford News* in July 1932:

> One day word came that sudden death had taken the mother of one of our little boys. Soon after, the father decided to move his family from Milford and the little fellow came for his books. I can see him now, a pathetic figure standing in the doorway, with spelling book, reader and slate under his arm, while we at the teacher's bidding all shouted in unison: "Good-bye Frank!" That boy was Frank J. Sprague, seven years old, just my age.[3]

North Adams was very different than Milford. Located in the far northwest corner of Massachusetts, North Adams, unlike shoreline Milford, was just about as far as one could be from the coast and still be in New England. The area was first laid out in 1739 by early settlers who saw the prospects for water power from the Hoosic River, which flowed through the town on its way to the Hudson River. North Adams was built in the low-level "notch" that carried the river northward through the town between the great Hoosac Mountain to the east and Saddle Mountain to the west.[4] The earliest construction of dams and mills began not long after 1750 along the Hoosic River just above the Main Street bridge. The first carding of wool into rolls began in 1804, and by the end of the Civil War the growing mill town was manufacturing and finishing a wide variety of wool and cotton material, while other manufactures included such industries as shoes, a blast furnace to make cast iron, an iron and brass

The big Beaver Dam on the North Branch of the Hoosac River supplied power to plants along Beaver Street east from North Adams' downtown.   *North Adams Historical Society.*

works, and a tannery. Typical New England mill buildings of brick or stone construction, often four or five floors in height, were erected along the river and its dams. By 1870 Adams had grown into a city of some 12,000 residents, with close to two-thirds of those living in North Adams.

In most respects North Adams was much like other New England mill towns of the time, but during the years that the two Sprague boys lived there it was also the site of one of the most ambitious American construction ventures of the nineteenth century. Construction of the Erie Canal had established a water connection from the Hudson River to the Great Lakes, and Massachusetts, anxious to establish its own connection to the west, had begun to think about developing a low water canal between Boston and the Hudson River. Loammi Baldwin, an engineer of internal improvements and canals, had completed a study in 1826, and had found by far the most favorable grades on the northern route across Massachusetts led through the passage along the Deerfield and Hoosic rivers, but this was blocked by the formidable Hoosac Mountain at North Adams. Nevertheless, Baldwin recommended the route and a tunnel as the best one to follow. The plan for a canal later became a plan for a railroad, and some preliminary work had begun on the nearly 5-mile-long Hoosac Tunnel in 1850, but the technology and equipment then available were inadequate to the task. The construction work was still underway and far from

Bird's Eye View from Kemp Park, North Adams, Mass.

Residential area of North Adams' Kemp Park looked outward to the lovely
Berkshire Hills of western Massachusetts. *North Adams Historical Society.*

complete almost 20 years later. By 1870, though, the work was finally making
headway, and North Adams was the center for tunneling workers, housing as
many as 700 men, many of them with their families as well, from the United
States, Canada, and Europe. New compressed air drills and nitroglycerin were
finally enabling progress to be made through the stubborn tunnel. A whole
factory for making nitroglycerin was set up at North Adams in 1868.

Not much is available to tell us about the lives of the two Sprague boys in
North Adams, but what there is suggests that they did quite well there. Anna
Sprague, later Mrs. Anna Parker, clearly took her responsibilities for the boys
seriously.

"She was a woman of the finest New England type and of striking beauty,"
Frank Sprague wrote of her years later.

> Living in a modest, frugal way, as an occasional school teacher, with great
> sacrifice she devoted herself to her charges with sanity of judgment, but
> with a high regard for much needed oversight. She was indeed a stern dis-
> ciplinarian, but I think that something vital must have been instilled in me
> by this devoted woman which race inheritance alone could not account for,
> something which was augmented by my later career in the Navy.[5]

Frank Sprague, in particular, seems to have been well known around North
Adams. In an article describing the young man's growing success (he was then

only 28 years old) in electricity, the writer for an August 1885 article for the North Adams *Transcript*[6] spoke of him as a schoolboy who was "bright-eyed, laughing and irrepressibly mischievous," and went on to describe his outgoing personality:

> He was a rollicking, good-natured chap. Constantly saying and doing provoking things which to a casual observer might indicate a careless, unambitious disposition, but to get offended at him or his pranks was impossible. His laugh would banish all feeling of irritation caused by his mischief.

The *Transcript* writer continued:

> His boyhood wasn't the most comfortable one in the world, so far as those things which come from ample resources are concerned, and he early learned the lesson on self-reliance. This was probably the most valuable training he received, for one of his strong personal characteristics is confidence in his own powers and dependence on his own exertions; and this is not inconsistent, either, with the fact he isn't afraid to ask for anything if he wants it, and is unconscious of objections or difficulties.

Both Frank and Charles attended the North Adams Public School and, later, Drury High School. Drury was originally established in 1840 as a private school, the Drury Academy, under a bequest from Nathan Drury, and was later established as a free high school. Frank Sprague proved to be a good student, particularly in mathematics. "Young Sprague attended the public schools here and was a remarkably bright, apt pupil"; according to the *Transcript* writer, "in fact, he was easily the smartest boy of his age in town."[7]

Knowing of Sprague's ability in science and mathematics, his high school principal had urged him to apply for the excellent free education provided by the Naval Academy or West Point. Sprague applied for what he thought was the examination for West Point, but when he arrived in Springfield, Massachusetts, in June 1874 to take it he found that it was for the Naval Academy. He took the competitive examination anyway, and stood highest among 13 candidates in the four-day examination. "A career afloat was far from my ambition," Sprague later recalled, "but having won out I decided to at least try it."[8] It would prove to be a fortunate one for Sprague, for there was probably no better choice than the Naval Academy for someone so strongly oriented to mathematics and physics.

Having done well in the competitive examination, Sprague was also recommended by such diverse figures as the North Adams probate judge; the pastors of both the Congregational and M. E. (Methodist Episcopal) churches;

Drury High School was originally founded in 1840 as a private school, but had become a public school by the time Frank Sprague attended it.   *North Adams Historical Society.*

and Walter Shanly, the contractor for the Hoosac Tunnel construction. "His uniform good scholarship, gentlemanly deportment and faithfulness in the performance of his duties have won for him the esteem of his teachers and associates," wrote Isaac W. Dunham, superintendent of schools and principal at Drury.[9] The 11th District, Massachusetts, Congressman Henry L. Dawes quickly recommended him for the appointment.

Frank and Charles had never enjoyed the close companionship of their father after he had left for the West in 1866, but there was at least some occasional contact. Learning of Frank's appointment to the Naval Academy, his father sent his warm congratulations in a letter he wrote from Denver on July 9, 1874:

> My Dear Son Frank,
>
> I received yours of June 23rd informing me of your success in getting appointed to the Naval Academy, and you can hardly imagine how glad I was to hear it, the more so that you got it without rich and influential friends to aid you, which some of the others undoubtedly had, I congratulate you heartily on your success, I wish your poor Mother was alive to be proud of her noble boy: but she is doubtless looking down from above with joy at your past and hope for your future success. If I had had the choice I could not have chosen a profession that would have pleased me better, and I hope and feel that you

have a very bright future before you, who can say but you may carve out a name in the country's history equal to a Perry or a Farragut.

Your Father, D. C. Sprague.[10]

To pay his expenses in getting to Annapolis, Sprague borrowed $400 from contractor Walter Shanly and a local bank, which he would carefully repay just as soon as he was able, and set out for Annapolis in September 1874. On his way, Frank Sprague got his first glimpse of the great city of New York, which would become his home for most of the rest of his life. Sprague wrote about this first visit many years later:

> I landed here in '74. The New York of that day was not the great metropolis of the present. There were no bridges, no river tunnels and no subways. There were no telephones, electric lights, no electric cars or elevators. Transportation was by horse-drawn streetcars or buses, while automobiles were still of the dream world. On the corner of 42nd Street and 5th Avenue was a great stone reservoir, and the vast territory running north of 72nd Street was largely barren and the home of goats and squatters.
>
> I little dreamed that I should ever in any way be a factor in the city's growth, but determined to make the most of this, my first visit, I climbed half-way up Trinity steeple to get a panoramic view of the city. Now that territory is occupied by a forest of skyscrapers, and all one can see from that vantage point would be across the cemetery.[11]

The story is told, too, of Sprague's great interest in architecture in New York City, as evidenced by his first look at St. Patrick's Cathedral, which was still under construction. Unable to gain access to the building from the workmen, he was told that only the architect or the Cardinal could grant it. Demonstrating what came to be his customary forthrightness, he promptly rang the bell for the attendant priest-secretary and asked to see the Cardinal. Cardinal McCloskey took an interest in the young man and quickly gave him a card that permitted him to roam through the cathedral as he wished.

On September 29, 1874, Frank Sprague successfully met the Naval Academy's requirements and accepted his appointment as Cadet Midshipman, and on October 3, 1874, he was sworn into the naval service.

The Main Gate of the Naval Academy in Annapolis, Maryland, greeted visitors in 1885.   *Special Collections and Archives, Nimitz Library, USNA.*

2

THE MIDSHIPMAN
INVENTOR

When Frank Sprague began his appointment as a midshipman at the Naval Academy, it was not, in some respects, the best time to be committing to a career with the United States Navy, for it was in the midst of a long period of decline. During the Civil War the Union Navy had built the greatest navy in America's history. At the end of the Civil War the navy had some 626 ships in commission—65 of them ironclads—but with the war won, and no threatening rival in sight, Congress was unwilling to support and maintain this great fleet. Its size steadily declined, with only 48 wooden hulled and obsolete vessels in service by 1880. Admiral David D. Porter, the navy's senior officer, compared them to "ancient Chinese forts on which dragons have been painted to frighten away the enemy." By 1878 the number of enlisted men had dropped to no more than 6,000, the lowest level in more than 40 years, and most of these were foreigners. There were not enough spaces for all of the officers who had graduated from the Naval Academy, and those that were assigned to ships had to wait considerable lengths of time for opportunities for new assignments or promotions. It was not until 1883, when Congress finally appropriated funds for the navy's first steel ships, that modernization of the antiquated fleet began.[1]

But if the larger navy was stuck in the doldrums, the post–Civil War Naval Academy in contrast was experiencing a remarkable period of change and growth that would be as great as any time in the nineteenth century. Expansion and improvement began in 1865 with the relocation of the Naval Academy back to Annapolis from its temporary Civil War location in Newport, Rhode

A view of the Severn River waterfront of the Naval Academy from the cupola of its New Quarters in 1873. *Special Collections and Archives, Nimitz Library, USNA.*

Island. The old buildings in Annapolis were refurbished and an extensive construction program for new buildings was begun. Rear Admiral David D. Porter, who had a long career and a brilliant Civil War record, was appointed as Superintendent of the Naval Academy in the fall of 1865.

Porter quickly began radical alterations to the academy. The curriculum was greatly modified, with additional emphasis being given to such topics as physics, history, mechanics, astronomy, English composition, and law. The old guard of professors had largely been replaced by a faculty of accomplished young officers who brought the experience of the Civil War to their teaching. The organization of the academy was almost completely modified, and Lieutenant Commander Stephen B. Luce was appointed Commandant of Midshipmen. Luce, a consummate seaman who was revered by the midshipmen who served under him, wrote the book *Seamanship,* which was the academy's text for the next 40 years, and later founded the Naval War College. Porter established an honor system, encouraged athletics, and established social activities. When he left the superintendent's post at the end of 1869, taking up President Grant's request to reorganize the Navy Department, the Naval Academy had been raised to an unprecedented peak of efficiency.

The Naval Academy's New Quarters housed the bulk of its midshipmen from 1869 to 1905. *Special Collections and Archives, Nimitz Library, USNA.*

Frank Sprague became a cadet midshipman in October 1874 at a time when the Naval Academy would begin yet another period of growth and improvement. Rear Admiral Christopher R. P. Rodgers, who had served as the Commandant of Midshipmen in 1861, took office as superintendent in September 1872. Rodgers took on the assignment with the aspiration of building on the work of his predecessors to bring the academy to a new standard of excellence, and he largely succeeded.

A longstanding area of dissatisfaction concerning the course of study and status of cadet-engineers was essentially resolved in 1874, when Congress abolished the two-year engineering program and established a full four-year program that shared many courses of study for the engineers with midshipmen. Cadet-engineers took advanced technical courses under a new Department of Mechanics and Applied Mathematics, and were required as a final test of proficiency to design and build a steam engine. This was the first course in mechanical engineering anywhere in the United States, and civilian colleges and universities sought to establish similar programs.

Admiral Rodgers revised the midshipmen's curriculum, with professional subjects in the first two years, adding upper level electives in mathematics, me-

Among the military training exercises assigned to all midshipmen was this infantry lead-ing drill, shown in the 1870s. *Special Collections and Archives, Nimitz Library, USNA.*

chanics, physics, and chemistry. The faculty of the academy during Sprague's undergraduate years included some exceptional teachers. Perhaps the most notable of these was Prussian-born Albert A. Michelson, who grew up in the mining camps of California and Nevada and graduated from the Naval Acad-emy in 1873 with high marks in such topics as optics, acoustics, and mathemat-ics. After his two years at sea, he returned to the academy as an instructor in physics and chemistry, at the same time beginning his work to measure the speed of light. Years later, Sprague classmate Vice Admiral Harry McL. P. Huse recalled Ensign Michelson's work on the velocity of light, remembering that he had

> rigged up some curious looking mirrors in one of the windows of Com-mander W. T. Sampson's house in Blake Row and other mirrors in a window of the Physics Department 200 to 300 yards distant. . . . We knew that he was seeking to measure the velocity of light through the deflection of a (light) ray by a revolving reflector, but we no more realized the far-reaching and immense importance of his work than we did the fore-shadowing results of the work of a youngster (Frank Sprague) in our own class who spent his recreation hours playing with gadgets in the Physics laboratory.[2]

Michelson went on to further study in Europe and later became head of physics at the University of Chicago. His award of the Nobel Prize in physics in 1907 was the first for any American.

Joining the academy faculty in 1874 were two remarkable seamen, Com-mander Winfield Scott Schley and Commander William Thomas Sampson,

The midshipmen's mess at the Naval Academy in 1887.
Special Collections and Archives, Nimitz Library, USNA.

who headed, respectively, the departments of modern language and physics and chemistry. Schley had been an assistant commandant of midshipmen and head of the French department, and would head a navy relief party for an army Arctic exploration in 1881. Sampson was a highly regarded teacher who had been an instructor twice before, and would later greatly advance the academy as its superintendent from 1886 to 1889. Sampson as a rear admiral and Schley as a commodore commanded the navy forces in the great victory over the Spanish fleet off Cuba in 1898.[3]

Frank Sprague's first two years at the academy were largely devoted to such basic academic subjects as mathematics, chemistry, physics, English, a foreign language, history, rhetoric, and drawing. The two senior years were concentrated on a wide variety of largely technical courses, such as marine engines, astronomy, navigation and surveying, applied mathematics and me-

Two midshipmen at study are shown in the Naval Academy's quarters in 1869.
Special Collections and Archives, Nimitz Library, USNA.

chanics, electricity, light and heat, composition and public law, and additional modern languages.[4] "There," Sprague recalled, "I developed something of a flair for mathematics, and particularly for naval architecture and physics, the latter under the teaching of that great admiral, William T. Sampson, one of the Navy's most brilliant officers."[5]

While the academy's academic requirements were demanding enough, the midshipmen were also expected to be well versed in a variety of naval subjects in seamanship, ordnance, and gunnery, and a substantial share of their schedule was devoted to the boat, sail, infantry, and light artillery drills required to sharpen their skills.

Together with the seamanship work that was made a part of the academy's daily life, each midshipman also participated in extended practice cruises over

The sloop of war USS *Constellation* served as a Naval Academy practice ship from 1872 to 1893. It is shown moored in the Severn River off the Naval Academy in 1879. The historic ship remains on display today in Baltimore, Maryland.   *Naval Historical Center (Neg. NH 61864).*

the summer period between academic years. Sprague made his first practice cruise aboard the historic practice ship USS *Constellation*,[6] which sailed from Annapolis Roads on June 26, 1875, on a three-month cruise along the Atlantic coast that included stops at Hampton Roads, Virginia; Buzzards Bay and New Bedford, Massachusetts; and Newport, Rhode Island. Led by Commandant of Cadets, Commander Edward Terry, the ship was headed by faculty officers and men from the academy, while 93 first-, second-, and third-class cadet-midshipmen manned the ship.[7] The midshipmen learned a seaman's work aboard ship, swinging their hammocks on the ship's berth deck, and carrying out both the duties of the ship's enlisted crew and study requirements. Frank Sprague began his final year at the academy with a second summer cruise aboard the USS *Constellation*, following much the same itinerary.[8]

By the time he had moved up into the Academy's third year, Frank Sprague was beginning to develop his social skills as well as his academic ones. For several years, beginning in about 1876, Sprague exchanged letters—as what we might call today a pen pal—with Mattie H. Munro, a young lady living in Boston. In one letter, Sprague reported to Mattie:

> Last Saturday evening the bachelor officers gave a "hop" (or dance). I thought it necessary the general welfare and happiness, especially of myself, to be there, and so I wandered down about 9:30 P.M. And now Mattie, for the crisis, I went on the floor, the first time I have ever ventured in a waltz with a lady at the Academy. I didn't fall, or step all over her dress, nor do any thing decidedly awkward.[9]

Sprague also told Mattie about the more advanced subjects—English composition, French, seamanship, astronomy, infantry tactics, and differential calculus—he was studying in his third year at the Academy, and told her of another technical subject he would soon be studying. "We soon have a course in practical electricity, that is, about nine of us," said Sprague with perhaps some foresight of his future. "I think I shall like it very much."[10]

Many years later a Sprague classmate, James H. Glennon—later a rear admiral—recalled an incident at the Naval Academy that told much about Frank Sprague's character. During the period from 1872 to 1874 the academy had admitted its first black midshipmen.[11] Upperclassmen had decided that a Negro cadet midshipman in Sprague's class would be given the "silent treatment." Sprague was interested in talking to this fellow midshipman and ignored the prohibition, and he was soon involved in a fistfight with a third classman over the matter. "You have not the pug nose of a fistic champion, [and] were a sorry sight after the battle," recalled Glennon, "but you licked your man."[12]

Sprague reached his graduation on June 20, 1878, with some notable achievements. Just getting there was one. Four years before no fewer than 103 cadet midshipmen had successfully passed the academy's competitive examination and otherwise met its rigorous requirements, but only 36—scarcely a third of them—made it to graduation, while another 14 cadet engineers were graduated. In his final year Sprague had been named a cadet-ensign for the academy's cadet-midshipman formation, and would stand at No. 7 in overall class standing upon graduation.

Many of Sprague's classmates in the Class of 1878 went on to notable careers. More than two-thirds of his classmates served long navy careers, with 13 of them reaching flag officer rank as commodores or admirals. Frank W. Bartlett, for example, served as an engineer officer on a wide variety of ships,

Naval Academy midshipmen joined a Class of 1873 musical group.
Special Collections and Archives, Nimitz Library, USNA.

including the battleship *Massachusetts* and the dynamite cruiser *Vesuvius*, in which he fought in the Spanish-American War in Cuban waters. He served in still other battleships and cruisers, was the fleet engineer of the Pacific Squadron, taught twice at Annapolis, and served for five years as inspector of engineering material before and during World War I. He was retired in 1920 as a commodore.[13]

Midshipman Frank Sprague during his Naval Academy years. *Frank Sprague Papers, Manuscripts and Archives Division, The New York Public Library, Astor, Lenox and Tilden Foundations.*

James H. Glennon saw extensive early experience in warships, and then served on the training ship *Constellation,* followed by two separate teaching assignments at the Naval Academy. He participated in the Spanish-American War on board the battleship *Massachusetts* in the Cuban campaign, and later commanded the gunboat *Yorktown* and the battleships *Virginia, Florida,* and *Wyoming.* Among his final assignments were a special mission to Russia during 1917 and the command of a series of Atlantic Fleet battleship divisions and of the naval district headquarters. He retired as a rear admiral in 1921.[14]

William Ledyard Rodgers came to the Class of 1878 from a long line of naval officers, with a grandfather who fought in the Revolutionary War, and a father who fought in the Civil War. Rodgers had extensive navy experience and served in the *Foote,* one of the Navy's first torpedo boats, in the Spanish American War in 1898. He served in several ships, taught at the Naval War College, and served as commanding officer on the *Wilmington* and battleship *Georgia.* During 1911–1915 he was president of the Naval War College, and during World War I he was given command of the supply ship crossing of the Atlantic. He later was appointed to the Navy's general board, as commander of the U.S. Asiatic Fleet, and on the Advisory Committee on the Limitation of Armament. He retired as a vice admiral in 1924. Rodgers also wrote extensively on military weapons and tactics, and two books, *Greek and Roman Naval Warfare* and *Naval Warfare Under Oars,* published after his retirement rank as important works on ancient naval warfare.[15] Sprague classmate Harry McL. P. Huse had both extensive service at sea and at the Naval Academy following his

The Naval Academy Class of 1878. The members have not been identified, but Frank Sprague is believed to be the man on the left, second row from
front. *Special Collections and Archives, Nimitz Library, USNA.*

graduation in 1878. During the 1898 Spanish American War he was executive
officer of the gunboat *Gloucester* in the Battle of Santiago and leading the party
ashore that first raised the U.S. flag over Puerto Rico. Over the next 15 years he
commanded a variety of ships, including the battleship *Vermont*. From 1914 to
1915 while serving as chief of staff for Rear Admiral Frank Fletcher, Huse was
awarded the Medal of Honor for his conduct during the landings at Vera Cruz,
Mexico, in April 1914. His later assignments included several senior navy posts,
and an overseas post as senior U.S. Navy representative on the Allied Naval
Armistice Commission. He retired in 1922 as a vice admiral.[16]

Frank Sprague also joined with other members of his class to go on to
notable careers in the civilian world. Mortimer E. Cooley, for example, who

graduated as a cadet engineer in 1878, served for more than a decade as navy engineer officer, including an assignment as chief engineer for the auxiliary cruiser *Yosemite* during the Spanish-American War, before going on to become the dean of engineering at the University of Michigan. Cooley served Michigan as dean from 1904 to 1928, and was regarded as the individual who guided the college's transition to the modern age of engineering. He is honored even today by his name on the modern building of Michigan's Department of Nuclear Engineering and Radiological Sciences.[17] Ira N. Hollis, who ranked first on the Naval Academy's listing of cadet-engineer graduates in 1878, soon moved to higher education as professor of mechanical engineering, and later overseer, of Harvard University from 1883 to 1913, and then went on to become president of Worcester Polytechnic Institute from 1913 to 1925. While at Harvard, Hollis designed and built the Soldiers Field stadium in Cambridge, Massachusetts.

The combination of Frank Sprague's inherent abilities, together with the strength and demanding academic requirements of the Naval Academy, had made him particularly well suited for the technical work that interested him. At the same time, the discipline and commitment that the Academy's military environment required would help him to work with the intense concentration and determination that would remain a Sprague characteristic for the rest of his life.

It was a time of great change and opportunity for young men like Sprague, who saw the many possibilities ahead in the exciting new field of electricity. In 1873 Zenobe Theophile Gramme demonstrated at a Vienna Exhibition how an electrical generator could be operated as a motor. In 1876 Alexander Graham Bell patented his new telephone. The first electric streetlights were installed in 1878. Thomas Edison applied for his first patent for an incandescent lamp in 1879, and large-scale electric lighting began just a year later. These and dozens of other new developments appeared in the rapidly changing electrical world.

In 1876 Midshipman Sprague journeyed to the great Centennial Exhibition in Philadelphia to see the wonders of new electrical and other ideas. Even as an undergraduate, Sprague had begun to develop and noted his innovative ideas for new applications of electricity, something he would continue for the rest of his life. Inspired by the works of Bell, Edison, and telephone inventor Elisha Gray, Sprague had developed a duplex telephone design. Seeking the use of some apparatus, Sprague—in his customary directness—had written to Edison shortly before his graduation from the academy. "The Western

Union Co. who own my patent would not allow me to do what you request," responded the young inventor. "If you could come here I would gladly give you every facility you require."[18] Sprague, on his way home from his Naval Academy graduation, did so. "Despite the fact that I was a stranger, a kindly reception immediately put me at ease, and a candid criticism, illustrated by a sketch of an alternative scheme, was emphasized when, to more fully satisfy me, I was told to go to the laboratory and experiment for myself,"[19] Sprague later recalled. It was the first meeting between Sprague and Edison, and the two would meet often for the remainder of their lives.

Frank Sprague had graduated from the Naval Academy, but he was by no means finished with the navy. At that time graduating midshipmen did not automatically advance to officer rank, but were continued as midshipmen for several years, and then appointed as ensigns after they had successfully passed a final examination. (This was finally changed in 1884, when a graduating midshipman was promoted to ensign as soon as he had successfully completed an examination.)

Late in 1878, Sprague was assigned to duty on the newly overhauled steam sloop *Richmond,* which was en route to become the flagship of the Asiatic Fleet. Launched at the Norfolk Navy Yard early in 1860, the *Richmond* was soon caught up in the Civil War, and was one of the ships in line at Mobile Bay when Admiral Farragut gave his famous command, "Damn the torpedoes . . . full speed ahead!" After the war, the ship served in European waters, on the West Indies Squadron, and then the South Pacific Station before beginning its Asiatic Squadron.[20] Sprague boarded the ship at Norfolk, Virginia, departing on January 11, 1879, to begin the long journey across the Atlantic, through the Mediterranean Sea and the Suez Canal, and over the Indian Ocean and North Pacific to finally hoist the flag of Rear Admiral Thomas H. Patterson at Yokohama, Japan, on July 4, 1879.

Sprague's work included a variety of naval duties. The ship's commanding officer, Captain A. Burham, was happy to report to the Secretary of the Navy at the time of his detachment back home to take his examination in 1880, that Midshipman Sprague "has performed his duty with zeal and intelligence." He had charge of the ship's deck when it was under sail, and of standing a watch in the engine room, under steam, and was left in both positions "as much as possible in his own resources." Sprague had gained considerable experience in navigation, said Burham, and "I should feel every confidence in trusting him with the navigation of a vessel."[21]

Despite the demands of his naval duties, Sprague found time for other pursuits as well. He managed to get himself assigned as a Special Correspondent

Early in 1879 Frank Sprague boarded the steam sloop *Richmond* to begin more than a year aboard the ship on its Asiatic Squadron.   *Naval Historical Center (Neg. NH 44997).*

for the *Boston Herald,* writing about his own experiences under the penname "Faix."

It was a time of peace, and the *Richmond* traveled around its Asiatic duties with pleasant visits to the seaboard cities of Japan, China, and the Philippines. General Ulysses S. Grant and his party, then on a long world tour after the end of his presidency in 1877, traveled for a time aboard the *Richmond* in the summer of 1879. General Grant traveled on the ship through the beautiful Inland Sea, and a little later, with the Mikado (Emperor) Mutsu Hito in presence, the Grant party and four accompanying *Richmond* midshipmen were invited to attend a grand race in Tokyo, followed by dinner with the general. "They evidently had a splendid time," recalls Frank Sprague of his shipmate friends, "I was of course on duty that day, and lost it all."[22]

But Sprague did make many of the magnificent social affairs of the Asiatic Squadron. In August 1879 he took several days of leave to travel in the Japanese countryside, and to make the celebrated climb up the 12,288-foot Fujiyama. One example of his social life comes from a letter from Sprague to Miss Frances Seale—"My dear Miss Frankie"—who was a regular correspondent with him during his Asian tour, on December 29 as the ship was preparing to depart to Hong Kong from a Christmas visit to Manila.

"But I must hasten to tell you of one of the most enjoyable affairs I was ever present at, and that is a ball at Manila," wrote Sprague. "We received an invitation to go last Saturday night (the day after Christmas) at ten o'clock. I was on the *qui vive,* and was determined to have a glorious time." The dance began and things moved slowly at first until Sprague and his shipmates began to understand how the protocol worked. The young ladies were almost all Spanish or Filipino, almost all of them dressed in striking *de sayo,* or local style, and few spoke English, but the midshipmen soon learned that you only needed to say "¿Quiere Ud. Hacerme el favor de bailar conmigo?" ("Will you do me the favor of dancing with me?") and the ball was underway. It took some help from Captain Burham to get Sprague going, but he was soon on his own. "After that I asked for myself, and until half past four the next morning, must I say it was Sunday, I did not miss an entire dance. . . . I never had so much fun in a strange place in my life."[23]

While he may have much enjoyed his social outings, Frank Sprague more than anything else wanted to get back home to the United States to join in the exciting new field of electricity. By the time he had completed his graduation, his creative urge had taken full possession, even while he was on duty on the *Richmond.*

> I was guilty of nearly three score of inventions of varied character, most of these are recorded in a much prized "Midshipman's Note Book," mixed with professional notes, sketches, and cruise records. A duplex telephone, pocket phonograph, time fuse, quadruplex and octoplex telegraph systems, a weird motor, means for transmitting pictures by wire, gyroscopic control of the mercury horizon and torpedo direction, an electric pantograph, a multiple telescope, regulation of incandescent lights, a water cooling and filterer apparatus, and control by variations of pressure on a submerged carbon disk of a ship's engines, to present racing with exposed propellers, are indicative of a variety of activities which were a nuisance to my shipmates. Many of these inventions were really worth while, but neither naval duties nor available money made possible their development then.[24]

Sprague was ordered back to the United States in March 1880, returning aboard the steamship *City of Peking* to San Francisco. Back home, he took and passed the final midshipman examination, although a promotion to ensign was not yet forthcoming. Briefly on leave, Sprague began some experimental work at the Brooklyn Navy Yard and Stevens Institute of Technology, in Hoboken, New Jersey, on an arc-lamp mechanism. While at Stevens, Sprague had met and talked with widely known inventor Professor Henry Draper (chemist, botanist, astronomer, and physician), and electrical inventors William Wallace, who was working with dynamos and arc lighting, and Professor Moses G. Farmer, who had demonstrated a small electric locomotive in 1847 and was now the government electrician for the Navy Torpedo Station at Newport, Rhode Island. For a 23-year-old fledgling inventor the work of these men was powerful encouragement and must have motivated Sprague as he began to develop his own inventions.

Sprague's short leave was soon over, and in the fall he was ordered to duty on board the *Minnesota,* an aging steam frigate then on duty at the Brooklyn Navy Yard as a gunnery and training ship for naval apprentices. Sprague was not enthusiastic about the prospective duties. "I found teaching the young idea how to shoot, reef sales and tie knots anything but agreeable work," he said, "and both at the Navy Yard in Brooklyn, and later at Newport, I improved every opportunity to put my ideas in metal."[25]

Sprague's first attempt was to put together a system of electric illumination on the *Minnesota* to replace the ship's crude system of oil lighting, operating an unused steam plant with a borrowed Edison "Z" dynamo. The project was a failure, however. Edison refused to loan the dynamo, pointing out that the proposed system would provide only a flickering lighting that would hardly advance the use of electric lighting on men-of-war.

The *Minnesota* was later ordered to Newport, where Sprague developed his relationship with Professor Farmer at the Torpedo Station. Led by Farmer, the Torpedo Station at the time had become a center for navy electrical development. Sprague gained approval to use the equipment in the machine shop, and was soon at work on new inventions.

"For three or four months," Sprague wrote in one of his frequent letters to Mattie Munro, "I have been trying to build a machine which is claimed to be an impossibility by most electricians, and have repeatedly failed, but intend to continue work here."[26] He did, and by the summer of 1881, he had successively built a double-wound armature, with internal field, the several circuits connected to a switch to give various series and parallel combinations, "which appears to be simple in construction and to promise quite efficient performances," remarked Professor Farmer.[27]

The Sprague design made two major departures from earlier electric motor design. The first of these involved the placement of the magnetic field. Up to this time, dynamos comprised an external magnetic field assembly, between

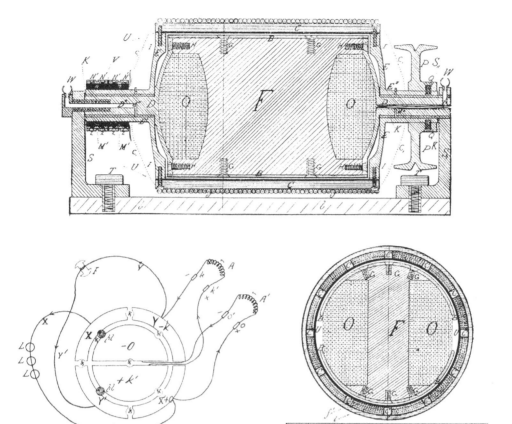

SPRAGUE INVERTED TYPE DYNAMO, SERIES-PARALLEL CONTROL, 1881

Working in Professor Moses G. Farmer electrical machine shop at the Torpedo Station, Sprague developed this notable dynamo electric machine which reversed the electrical field and armature of a dynamo, and featured an arrangement of field and armature circuits that became a basic principle of all series-parallel controllers used on DC railway motors.   *Middleton Collection.*

the poles of which the armature was rotated. In Sprague's new machine these relations were reversed, the armature being turned inside out and the coils enclosed by an outside iron shell of iron wire and inwardly projecting ribs, the whole surrounding the field magnet. Built for continuous current and with two armature circuits and commutators, the field magnet was held stationary, while the armature was rotated. This arrangement with external armature and internal field became characteristic of modern power plant alternators, in which the armature, or "stator," enclosed the field magnet, or "rotor," which became the moving part. The second innovative feature was a switch that enabled different combinations of field and armature circuits, a basic principle of all series-parallel controllers used on DC railway motors.[28]

Aware of the great French Electrical Exhibition in Paris planned for the fall of 1881, Midshipman Sprague asked to be sent as an assistant to the officer who would make the trip, but was refused. Sprague then tried to get an assignment to a ship going to Europe, together with three-months' leave on arrival in Europe. Moses Farmer wrote to the Secretary of the Navy, urging his approval, and Sprague was ordered to temporary duty on board the *Lancaster,* en route to Europe to take up the assignment of flagship for the Mediterranean Squadron. The ship was delayed for several months, but Sprague used the extra time to install a system of electric bells for the ship. "Candor compels me," he said years later, "to admit that neither material nor workmanship was up to modern standards, and before long it was a question whether the captain was calling the first lieutenant or the cook."[29]

The *Lancaster* finally got underway from Portsmouth, New Hampshire, in September and arrived at Gibraltar on November 9th. By this time the Electrical Exhibition in Paris had already ended, but Frank Sprague was not to be denied. Learning that another Electrical Exhibition would open early in 1882 in the Crystal Palace in Sydenham, England, near London, he had soon obtained orders to go there instead.

The youthful Frank Sprague—who finally was advanced to ensign on March 10, 1882—was made a member of the Jury of Awards, asking to serve on dynamo-electric machinery, where he came into contact with a number of notable scientists.

> Among my confreres were many men of science whose names have become of world renown, among them Prof. Fleming Jenkin, of Edinburgh University, inventor of telpherage; Capt. Abney, the great photographic expert; Prof. [W. Gryll] Adams, of the Wheatstone Laboratory, King's College, brother of Charles [Couch] Adams, one of the mathematical discoverers of [the planet] Neptune; Horace Darwin, son of the great naturalist, Charles Darwin; Prof. Frankland, C. E. Spagnoletti and others.[30]

Sprague—the youngest member present—was made secretary of the scientific group.

Frank Sprague organized a series of tests of dynamos, incandescent lamps, and gas engines that were said to be the most comprehensive ever undertaken at the time. The topics covered in the exhibition included descriptions and tests of gas engines, dynamo-electric machines, arc lights, and incandescent lamps and systems of distribution, with equipment supplied from a number of companies in both Europe and the United States. The navy had expected some sort of report on the exhibition, but Sprague went far beyond this with a comprehensive study of the exhibition that included detailed discussions of all the equipment and the testing results, figures and charts, and a number of detailed hand illustrations.

A particularly interesting area of Sprague's work was the testing of a variety of gas engines. One of those tested by Sprague was one that was operated without outside ignition, perhaps the first demonstration of the principle later developed by Dr. Rudolf Diesel for the Diesel engine in the late 1890s.[31]

With all of the work in testing and reporting that Sprague had taken on, his stay in London exceeded by several months what the navy had permitted him, and he proceeded with what he himself called a "liberal interpretation of my orders." Later, he wrote "it was with something of a shock that I received a sharp reminder from the Navy Department, with imperative orders to at once rejoin my ship at Naples, where I went with visions of court-martial and possible ultimate disgrace."[32] Sprague's version of this in later years was probably written with some exaggeration. A letter from the Bureau of Navigation on December 11, 1882, said simply, "Your letter of the 28th, respecting the reasons for not reporting for duty on the European Station is received, and are entirely satisfactory." You can remain there until your work is complete, said the Chief of Bureau, and please let us know when the report might be finished. In any case, Sprague was running out of money, and soon went ahead to the *Lancaster* to finish the report. The elaborate 169-page report, *Report on the Exhibits at the Crystal Palace Electrical Exhibition, 1882,*[33] published by the Navy's Bureau of Navigation, Office of Naval Intelligence, in 1883, attracted wide commendation.

During his year in London, Sprague also rode on the Metropolitan District Railway, the city's pioneer 4-mile-long underground railway. Opened in 1863, the underground was a success, but the smoke, gases, steam, and heat from its operation with coal-fired steam locomotives in a confined space did not make a Metropolitan journey a pleasant one. Traveling on the line regularly, and well aware of its deficiencies, Sprague soon developed his ideas for a much more satisfactory electric operation of the line. As he envisioned it, electric underground trains would operate in two planes, making upper and lower contact with them, with these two planes representing the termini of a constant potential system of distribution. One would be made up of the running tracks, yards, and switches, and the other by a center overhead rail following the center lines of all track and switches, with contact being made on the lower plane by the running wheels, and on the upper plane by a universal spring–supported de-

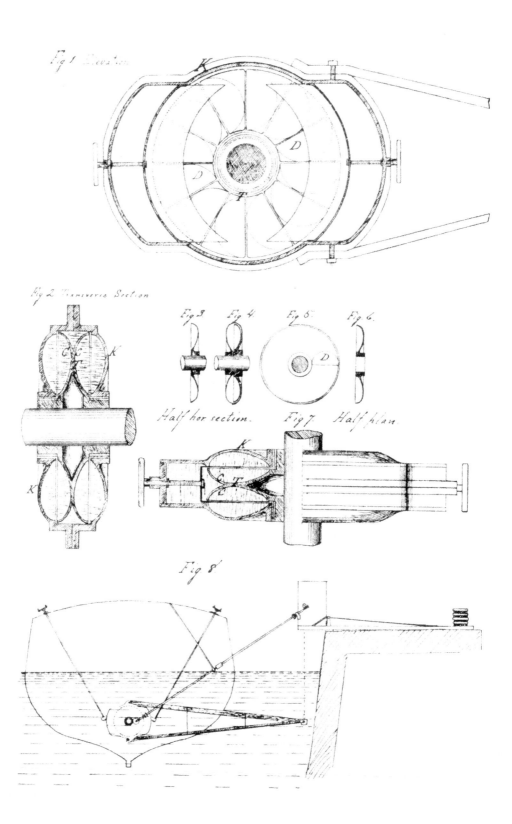

Water dynamometer developed for the testing of Otto and Clark engines. Drawing from Frank Sprague's report on the Crystal Palace Electrical Exhibition.   *Library of Congress.*

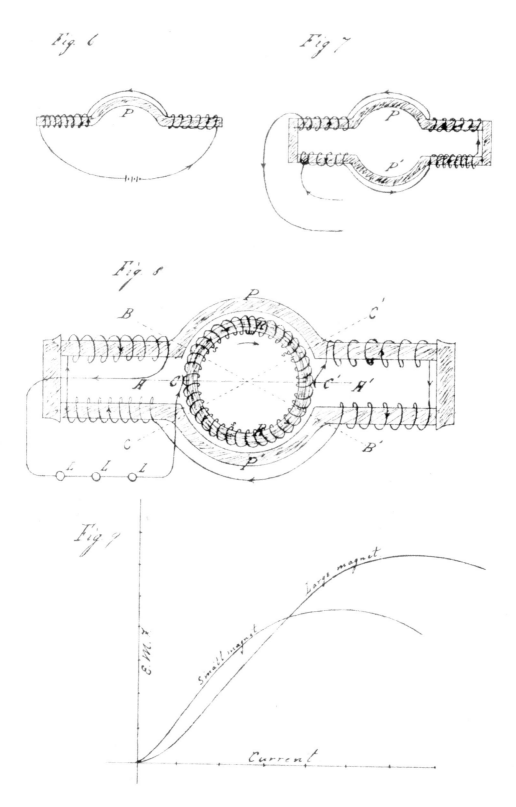

Gramme auto-exciting dynamo. Drawing from Frank Sprague's report on the Crystal Palace Electrical Exhibition. *Library of Congress.*

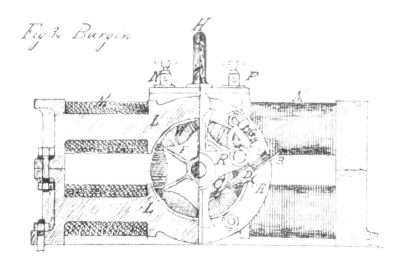

Fig 3. Burgin

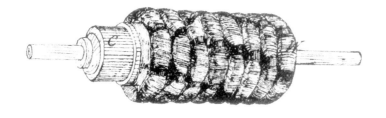

Fig 37 Armature

Fig 38 Diagram of Connections

vice.[34] At the time, Sprague did not go beyond these conceptual ideas, but—as always—he retained the ideas, and only a few years later electric operation of urban transportation would become one of his major objectives.

If there had been any doubt about the future direction that Frank Sprague might take, it was clearly resolved by his experiences at the Crystal Palace Electrical Exhibition. His future would lie with electricity. As his European stay drew to a close, Sprague had already submitted his resignation to the navy and taken a position in electrical lighting work with Thomas Edison.

Bürgin electric motor. Drawing from Frank Sprague's report on the Crystal Palace Electrical Exhibition. *Library of Congress.*

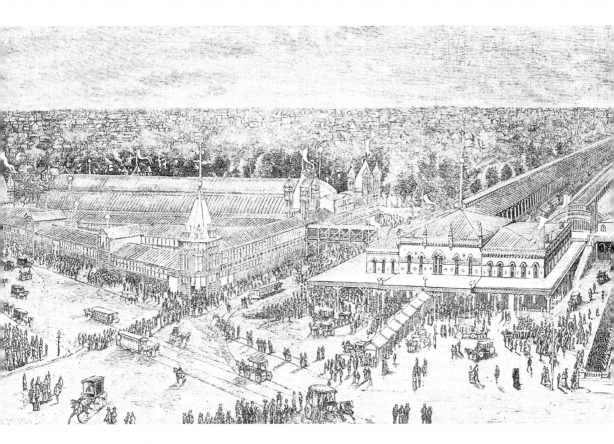

The 1884 International Electrical Exhibition in Philadelphia where Frank Sprague received much favorable mention for his newly developed constant speed electric motors.   Journal of the Franklin Institute, *from the Historical and Interpretive Collections of the Franklin Institute, Philadephia, Pa.*

3

SPRAGUE AND THE NEW WORLD OF ELECTRICITY

"A course of study which I have followed for four years has very strongly developed my tastes for work in connection with electrical service, and I can only feel satisfied when thus employed," wrote Ensign Sprague in a March 1883 letter to Secretary of the Navy William E. Chandler resigning his commission. Among other reasons Sprague cited for his resignation were his desire to engage in experimental work, and the receipt of attractive offers from several companies. The problems of the overcrowded condition of officers in the naval service and the slowness of promotion in the antiquated and under-funded navy also strengthened his desire to seek a career for himself in civil pursuits.[1] In any case the navy agreed, giving Sprague a year on leave, with his resignation to become effective April 15, 1884.

Sprague was engaged in work at the Crystal Palace Exhibition in 1882 when he became acquainted with Edward H. Johnson, an electrical engineer and inventor and a close associate of Thomas Edison, who would work closely with Sprague off and on for the next 15 years.

Johnson, who was then 36 years old, had begun his work as a Pennsylvania Railroad telegrapher, and then worked in telegraphy with other railroads and the Western Union Company. In 1866 he became telegraph constructor to General William J. Palmer, who was then engaged in the building of the Kansas Pacific Railroad, and later the Denver & Rio Grande Railroad. Johnson became an assistant to Palmer in the construction of the Rio Grande. Both Palmer and Johnson became interested in the acquisition of the Automatic Telegraph Company, and Johnson became acquainted with Edison in his work with Automatic Telegraph. Johnson and Edison collaborated on improvements to a new automatic telegraph machine developed by George D. Little, and Johnson was soon involved in a number of Edison's many projects, becoming the general manager for one of Edison's companies in 1876. A year later he was put in charge of the exhibition of the Edison telephone in the East, and in 1878 also exhibited Edison's phonograph and Alexander Graham Bell's telephone, and became general manager for the Edison Speaking Phonograph Company. In still other tasks for Edison, Johnson was involved in work on electric lighting and the manufacture of phonographs.

In addition to these many occupations in the United States, Johnson also traveled to England for Edison. The first trip, in 1879, was in the role of chief engineer of the Edison Telephone Company, amalgamating the Edison companies in Great Britain into the United Telephone Company. Johnson returned to England in 1882 to handle the installation of incandescent lights in the Crystal Palace, and then the formation in London and other English cities of the Edison Electric Light Company. In subsequent years Johnson would be engaged in a number of industrial firms, including both Edison companies and others.[2]

Sprague had been much impressed with the Edison equipment displayed at the Crystal Palace Exhibition. "In the entire exhibition there was nothing which so impressed me as Edison's work," Sprague recalled, "and in connection with this I was brought into contact with E. H. Johnson, whose buoyant belief in the work of his principal, coupled with my admiration for what had been accomplished, made me an ardent convert to the Edison system."[3] Johnson was evidently impressed with Sprague's potential, and before leaving London for the United States in June 1882, Johnson had offered him a job with Edison.

Before returning home, Johnson had sent Sprague to Manchester to examine a new design by British electrical engineer Dr. John Hopkinson for a three-wire design for lighting circuits, comparing it to a circuit developed independently by Edison.

The basic idea of Edison's three-wire system was that it made possible the use of a higher voltage, which in turn reduced the current requirement and the size of electrical wiring. Doubling the voltage reduced the current requirement by half for the same power delivered to the load. This in turn reduced the power lost in the distribution system by four in the same size wire, meaning that a wire one-fourth the size would have the same efficiency with the

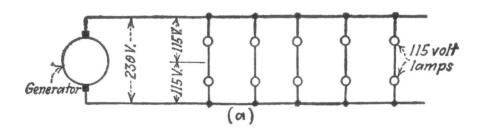

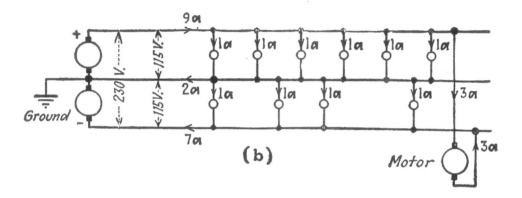

Thomas Edison's development of the three-wire system for lighting circuits was a major development in reducing the costs of electrification. Drawing (a) shows a typical two-wire circuit, which required two 110-volt lights in a series to employ a higher voltage 220-volt system. Drawing (b) shows how the addition of a third neutral wire made possible the use of 110-volt lights individually. On the left is shown the arrangement using two 110-volt generators, and on the right the arrangement with a single 220-volt generator.   *Middleton Collection.*

higher voltage. Already, however, incandescent lamps were being produced for a nominal voltage of 120 volts, while doubling the voltage to 220 would require a longer and more fragile filament, making the lamps harder to produce. One way to overcome this problem would be to simply connect two 110-volt lamps in series on a 220-volt wiring, but this would prevent the lamps from operating independently; one burned-out lamp would cause the other one to go out, too.

An Edison three-wire system used a potential of 220 volts wired between two wires, with a third, neutral wire between them that was at a potential midway between the two. This would provide a potential of 110 volts between the neutral wire and either the positive or negative wire, allowing a 110-volt lamp to be used on either side. Careful balancing of the loads between the positive and negative wires would then balance the currents on the neutral wire so that it would carry almost no current, allowing still further reductions in the wiring required to carry the load. Ultimately, Edison received a U.S. patent for the three-wire system, while the system was patented jointly by Edison and Hopkinson in Great Britain.[4]

Sprague finally left for the United States in the spring of 1883, arriving in New York on May 24, 1883, the day that the Brooklyn Bridge was opened. Sprague promptly reported to his new employer, who, Sprague later recalled, seemed to think that the salary of $2,500 agreed to for him was unduly munificent. Johnson had recommended that Sprague be hired as an expert on electric railways, but Edison instead hired him as an electrical expert for the new construction department which he had set up as an independent entity to promote and install central stations using the Edison lighting system.

Frank Sprague and William S. Andrews were sent to Sunbury, Pennsylvania, to install both the first station built by the construction department and the first overhead three-wire system. Andrews had joined Edison at Menlo Park, New Jersey, in 1879, where he had worked in the machine works testing department and had supervised the first tests of a small-scale version of Edison's three-wire system. Edison himself supervised all the technical work and approved all decisions concerning station design, and would personally supervise the Sunbury station because of its experimental status.[5] Sprague and Andrews were sent to Sunbury with instructions to complete preparations in 48 hours, and the three-wire plant was ready for test early in July.

"To me getting ready meant trial as well, and so the night before the Fourth of July an Armington-Sims high-speed engine was started and current was delivered on the line," Sprague later recalled. He continued:

> Sight feed oil cups were then something of a novelty, and having run some hours with little or no oil, we managed to burn out the babbitts, and despite diligent scraping by a local machinist, Mr. Edison when he arrived the next morning with his secretary and chief engineer found a badly pounding engine. I pass by the comments excited by my assumption of responsibility, but the plant continued to run.[6]

The basic purpose of the three-wire system of electrical distribution, as opposed to the two-wire system it replaced, was to reduce the amount of copper wire required in incandescent lighting circuits. Originally, explained Phillip A. Lange,

> [a] huge map was prepared, showing the location of the streets and the position of the houses where current was to be supplied. On this map, a spool of German silver wire was located wherever a house was to be supplied with

lights. Each spool had a resistance proportional to the resistance of the lamps in the house. Wires corresponding to the feeders to be actually used were stretched along the streets, and the German silver spools were connected to these wires. Current was obtained from a small Daniel battery, and distributed to the different spools through the wires. A professor then sat in front of the map and measured with a galvanometer the drop along each of the wires. From his measurement, the proper wires for running along the streets of the city could be determined.[7]

This was a typical Edison response to a problem, to use an experimental approach for determining or obtaining a required result. Edison and most of his associates at the time (1883) did not have the benefit of a technical education and mathematical training. When asked to look at the problem, Frank Sprague quickly saw that the model method was not only time consuming and expensive but also inaccurate, and he developed mathematical methods by which the correct sizes could be determined by calculation alone. "He quickly showed how to calculate the drop in feeders without laying out a whole city in miniature," Lange wrote years later, "determining in a few hours or minutes results which had previously required weeks of experimental work and a considerable financial outlay." Sprague applied for a patent on the system he had designed on September 19, 1885, and assigned it to the Edison Company.

Following completion of the Sunbury plant, Sprague went on to complete construction of the plant at Brockton, Massachusetts, Edison's first underground three-wire distribution system. Brockton was finished by October 1, 1883, but Sprague stayed on as operating engineer for the plant. In his spare time, Sprague continued to experiment on motors, and built an early electric railway motor.

Sprague was not much interested in working on electric illumination, which was Edison's chief concern at the time. Electricity for power, Sprague believed, would become equally, or even more, important than electric illumination. By the spring of 1884, Sprague was ready to move from Edison's lighting business to power generation. While he was pessimistic about Edison's developing much interest in power, Sprague did make an effort to continue working with Edison, but on his own terms.

In later years Sprague sometimes joked about his "being fired by Edison," but it wasn't quite that simple. Edison had asked Sprague to take up some problems associated with power transmission. Sprague wrote to Edison on April 24, 1884, discussing his great interest in developing the problems of electric power transmission. He would like to do this work, he said, but not under the present arrangement.

I feel that I can, and to a great extent in my own mind have, solved the question of this transmission. To take up this subject in obedience to your request would simply be to take over my own work without due consideration, and a proper regard for my future makes it impossible for me to do this·

As your subordinate, I cannot work with the same freedom as if I take the future into my own hands. Personal reasons, and my relations with others make it necessary that I should look well to the future, and with the confidence I feel, and the example of your own perseverance, I am willing to take upon myself whatever responsibility attaches to my action.

Sprague offered his resignation if Edison desired, but said: "Should it be desirable that I continue any relation with your work, I can only consent to do so in a purely consulting capacity, with a perfect freedom to the time and title of my own inventions."[8]

Edison wasted no time in considering the Sprague proposal. "As we are about to close out our construction dept. I think the best way is for you to resign on the 1st for the reason that your position would be so curious as to be untenable," Edison replied later the same day.[9]

This is doubtless what Sprague expected.

Sprague continued to work on his own on the development of electric motors, and by that fall he was ready to begin marketing his stationary motors for industrial purposes. This was just in time to show the motors at a new exhibition that was opening in Philadelphia. The Franklin Institute in Philadelphia had organized an International Electrical Exhibition unlike anything that had ever been seen in the United States, which would open on September 2. Encompassing more than 45,000 square feet of exhibition space, the exhibits were grouped by electrical production, conductors, measurement, and power applications, as well as material covering terrestrial physics, historical apparatus, and educational and biographical material. In addition to the electrical exhibits, the brilliantly illuminated show included music from the electrical Roosevelt organ and the Germania Orchestra, and periodic scientific lectures. The popular exhibition had drawn some 300,000 visitors by the time it closed on October 11.

Sprague's motors displayed at the exhibition were well designed, incorporating several new features, and their merits were quickly recognized. The motor was virtually non-sparking, and operated at a constant speed, regardless of changes in load. The motors were a revelation for such notable electrical men as British-born physicist Silvanus P. Thompson and electrical engineer Sir William H. Preece, or Thomas Edison himself.

Edison was a regular visitor to the exhibition, and was quite enthusiastic about the Sprague motors in a meeting with a reporter for the *Philadelphia Press.*

"And the transmission of electrical force?" asked the reporter.

"That problem has been pretty well worked out," said Edison of his recently departed electrical assistant.

> A young man named Sprague, who resigned his position as an officer in our navy to devote himself to electrical studies, has worked this matter up in a very remarkable way. His is the only true motor, the others are but dynamo-machines turned into motors. His machine keeps at the same rate of speed all the time and does not vary with the amount of work done, as the others do.[10]

Sprague wrote later:

> Our initial industrial motor development was based upon the important fact that on a constant potential circuit the mechanical effects—variations of speed and power output—of a motor could be controlled by inverse variation of the strength of the magnetic field to determine the differential of the line and motor electromotive forces; also, that with two magnetizing field coils, one of high resistance across the line for the main field excitation and another of few turns in opposition to it and in series with the armature, it would be possible with certain proportions to operate a motor at the same speed under varying loads, and even different potential differences, which constant speed might also be varied—this of course a mathematical deduction. In addition, it appeared that by a distorted location of the series coil it would be possible to maintain automatically a fixed non-sparking position of the brushes under varying loads. To insure a strong field in starting, a cut-out or reversing switch was added.[11]

This had been the genesis of what was known as the constant speed motor (what is today called a compound-wound motor) with fixed non-sparking position of the brushes, primarily for use on constant potential circuits. It was the first motor of the type to be put into commercial service. From the original principles developed by Sprague was the idea of regeneration of energy to return power to the supply circuit for braking of trains or elevator operation.

Sprague had also gone ahead with formation of the Sprague Electric Railway & Motor Company, which was incorporated late in November 1884. He was joined by Edward Johnson and John C. Tomlinson, an attorney for Edison, as the company's trustees. Johnson was president of the company, and Sprague did about everything else. "I became the electrician, office boy, treasurer, mechanic and administration man," Sprague later recalled.[12]

> The company was really a paper one with a nominal $100,000 capital, all of which was issued to me for inventions and patents, and for which I was to perform sundry services for a salary of $2,500, which I was to pay to myself. A few shares having been sold at par to a professional friend and soon used for personal expenses, I made a verbal contract with E. H. Johnson, then president of one of the Edison Companies, by means of which he was to advance certain moneys for a specified interest in the company. One small room sufficed for our business requirements, and the stationary motor development was carried on in the shops of Bergman & Company [a manufacturer affiliated with Edison], New York.[13]

Bergman was merged with other Edison companies in 1899 as Edison General Electric.

Production of the motors began immediately after the successful exhibition at the Philadelphia Exposition, but for several years the Sprague Electric Railway & Motor Co. remained little more than the modest company it had started as. While Sprague had ended his employment with Edison, he remained in close association with the Edison companies. Manufacturing work for the Sprague Company was all subcontracted to the Edison Machine Works, so Sprague manufacturing staff and equipment were not necessary, while the sales work for Sprague motors was handled by independent sales agents who invested their own funds in the inventory they carried.

The principal markets for the Sprague motors were firms which could obtain power from a central power station or their own incandescent lighting generators, principally those developed by the Edison constant potential circuits. This could be an attractive business, since electric current for illumination was largely required in the nighttime hours, while power for manufacturing was usually used in the daytime, making the cost of the additional power during the daytime hours quite low. Recognizing that the addition of a motor load could be quite profitable, the Edison Electric Light Co. published a circular in May 1885 to its licensee stations encouraging the development of motor loads, using Sprague motors for the loads.

"A practical motor has been a want seriously felt in our system," said the Edison circular, "and the value of it as a consumer of electric current, especially during hours of daylight, when the maximum of current is not required for lighting purposes, can be easily appreciated.

Completed in 1884, this small electric motor by Frank Sprague is still in operable condition at the permanent exhibit "Frank J. Sprague: Inventor, Scientist, Engineer" at the Shore Line Trolley Museum in East Haven, Connecticut. *William D. Middleton.*

"The Sprague Motor is believed to meet to the fullest degree all the exigencies of the case, and the Edison Electric Light Company feels that it can safely recommend it to its licensees as the only practical and economic Motor existing today."

Sprague anticipated widespread application of electric power. One early Sprague booklet listed these five uses of electric power: (1) the transmission of power over considerable distances; (2) the distribution of power within a single plant; (3) auxiliary power; (4) portable apparatus for pumping, hoisting, and so forth; and (5) tramways for the transportation of freight. Sales of the Sprague motors were brisk over the next several years for use in clothing factories, printing presses, and various kinds of mills. A Sprague motor was used in 1886 to operate a freight elevator in a six-story building. A Boston furniture dealer installed a Sprague motor–powered passenger elevator. In addition to being president of the Sprague firm, Edward H. Johnson also happened to

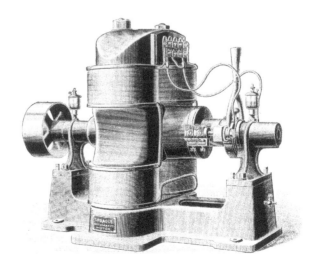

This larger stationary Sprague Electric Railway & Motor Co. electric motor, illustrated in an 1886 publication, used a rheostat to start the motor, and was normally operated at about 100 volts.   The Electrical World.

head several other companies, one of which was as president of the Boston Edison Company. This was probably helpful with the sale of Sprague motors around Boston, and by the end of 1887 there were 73 customers receiving electric power from the Boston station, operating a total of 240 horsepower.

By the beginning of 1887, there were 250 Sprague motors in service in the United States, ranging from ½ to 15 horsepower, and in the company's earliest catalogs the number, owner, locality, and duty, followed by statements from the users, identified every motor which had been put into use. By 1888 a catalog listed a total of 16 different Sprague motor sizes, with motors all the way from ½ horsepower to 100 horsepower. By this time, Sprague had decided to establish his own manufacturing plant, and a New York factory was leased in October 1886. To finance this, the company increased its authorized capital stock to $1,000,000. The new plant was used to build motors for special purposes, while the Edison plant continued to be used for manufacturing standard motors.[14]

Sprague motors were widely used for industrial purposes, and sold well, but once the problems of developing and manufacturing the motor had been solved, Frank Sprague began to give an increasing share of his time to a new problem that could be solved by electricity, the urban rapid transit railway.

A NEW FAMILY

Busy as he was with both his rapidly growing electric motor business and his work on developing electric railways, Sprague nonetheless found time for other interests as well. He took a short vacation trip to New Orleans in the spring of 1885. While there he met and was captivated by a beautiful young woman, 21-year-old Mary Keatinge, six years Sprague's junior and still recovering from a brief failed marriage. Mary's father, Edward C. Keatinge, an artist and steel engraver, had died soon after the Civil War. Her mother, Harriette Charlotte Keatinge, M.D., had taken up medicine after her husband's death and had become famous as the pioneer woman physician of the Gulf States. After a whirlwind courtship, Frank and Mary were married at an evening ceremony at New Orleans' Trinity Church, on April 21, 1885, followed by an elegant supper at the Hotel Royal. "This marriage," lamented the Society column of *The Daily Picayune,* "will remove permanently from New Orleans one of the loveliest and most charming girls." After a brief honeymoon, the couple returned to New York, where Sprague was soon again buried in his work, while Mary tried to create some kind of a social life.

Their only child, Frank D'Esmonde (later spelled Desmond) Sprague, was born in New York on March 29, 1888, with Sprague's mother-in-law, Dr. Har-

On April 21, 1885, Frank Sprague married Mary Keatinge in New Orleans. Although Mary and Frank divorced in 1895, their son, Frank D'Esmonde Sprague, would work with his father throughout his life.   *Courtesy of John L. Sprague.*

riette Keatinge, acting as attending physician. Frank was close to his son, and Desmond often traveled with his father when he was young. Desmond graduated from Cornell University as a civil engineer in 1911, and would work for and with his father throughout the time of his work with the Naval Consulting Board during the Great War and afterward with the Sprague Safety Control & Signaling Corp. until almost the time of Frank's death in 1934.

The marriage between Frank and Mary, however, turned out to be disappointing for the couple. Frank was heavily involved in his electrical work, which was little understood by and of little interest to Mary, while her social environment was of no interest to Frank. The couple slowly drifted apart and they were divorced amicably in 1895.

But even though divorced, they and future spouses would continue to be involved in a cordial relationship throughout their lives. Mary was later married to Anthony Morse, and then very happily with a distinguished Hindu scholar and professor, Dr. Taraknath Das. Dr. Das had become a naturalized U.S. citizen, only to lose it when the U.S. Supreme Court decided that Hindus were not of the white race, and both Das and Mary lost their citizenship. The Das family often found themselves in difficult financial straits, and the Spragues (Frank and his second wife Harriet) often provided financial help. The relationship continued long after Frank Sprague's death. As recently as 1947, Mary Das wrote to Harriet Sprague, congratulating her on the publication of her book, *Frank J. Sprague and the Edison Myth,* and remembering Frank Sprague as "the kindest, most brilliant man I ever knew."[15]

ELECTRICITY FOR THE ELEVATED RAILWAYS

By 1880 the elevated railroads in New York were running close to 2,000 daily trains and transporting 61 million annual passengers. Construction of elevated railroads in Manhattan had largely been completed in 1880, but work on the elevated lines in Brooklyn was just beginning in 1885, and within another decade Chicago would begin the construction of an elevated system that would be second only to that of New York. Steam operation of the elevated lines, however, left much to be desired. Residents and businesses along the elevated routes objected to the steady rain of smoke, cinders, and steam from the locomotives, while hot coals and sparks dropped into the streets below. Steam operation, too, was uneconomical and inherently limited in its performance characteristics for the stop-and-go nature of its operation.

An early elevated rail line in Manhattan used a continuous cable–powered system which opened in 1868, but its frequent problems soon brought the

Steam-powered elevated railways made possible New York's growth in the years after the Civil War. But smoke, cinders, and noise made them unpopular. The New York Elevated Railroad is shown here in 1879. Frank Sprague and other electrical inventors tried to convince the elevateds (an abbreviated name that gave rise to other nicknames, such as the "El" in New York or the "L" in Chicaco) to convert to electric power. It would take almost 20 years before they finally did.   *Library of Congress.*

line to a close, and it was replaced with steam locomotives. By 1880, however, the rapidly developing technology of electrification had begun to show some promise as an alternate power source for railways. In 1879 German inventor and electric engineer Ernst Werner von Siemens operated the first electric railway powered from a dynamo, or generator, followed over the next several years by similar tests by Stephen D. Field and Thomas A. Edison with electric locomotives. Early electric street railways were also being developed by Leo Daft, John C. Henry, Edward M. Bentley, Walter H. Knight, Sidney Short, and Charles J. Van Depoele beginning from about 1882.[16]

To Frank Sprague and other early electric railway inventors, the development of electric power looked like a promising replacement for the elevated railways' steam power. In 1884 officials of the Manhattan Railway were interested enough in the possibilities of electric traction to want to take a close look at its capabilities. The elevated railway men, together with representatives of

Dr. Ernst Werner von Siemens was an important early developer of the electric railway. This passenger carrying line, built for the Berlin Industrial Exhibition in 1879, was the first dynamo- (generator-) powered railway ever built.   *Siemens A. G.*

the Siemens, Field, Edison, Daft, Brush, and Bentley-Knight electric railway companies, met jointly in New York on November 4, 1884. Through this and subsequent meetings it was decided to test each system on an elevated railway track, with a commission appointed (among them Scottish physicist Sir William Thompson) to test the motors and adopt the best system.

The elaborate program of tests failed to materialize, but two of the companies would go ahead with test operations. The most important of these was by Leo Daft's company, the Daft Electric Light Company. Beginning early in 1885, the company put in a center third rail for electric power over a line about 2 miles long on the double track Ninth Avenue elevated line between 14th and 53rd streets, with separate third rails being installed on the uptown and downtown lines. A power plant consisting of a Wright steam engine and three dynamos was installed at 15th Street, about 250 feet west of the elevated tracks. Daft completed a 9-ton electric locomotive, the *Benjamin Franklin,* to run the Manhattan elevated tests. The four-wheel locomotive was equipped with a 75-horsepower motor powered through a friction arrangement to one of the two axles, and equipped with a large electro-magnet brake. The power supply used a 280-volt DC third rail line.

The *Benjamin Franklin* was ready for operation on August 26, 1885, and successfully powered two to four standard elevated cars over the Ninth Avenue line. The *Franklin* experienced some problems with its friction drive, and Leo Daft became convinced that the locomotive was too light to carry the heavy trains of the elevated. The locomotive was taken out of service late in 1885 and rebuilt with a geared drive that powered both axles of the *Franklin*. Back in service late in 1886, the *Franklin* still simply could not pull the heavy four-car trains that were standard on the elevated, and the locomotive was again taken out of service.

Daft made a third try at meeting the elevated's carrying capacity requirements with a much improved and reconstructed *Benjamin Franklin* that was back in service and tested during October and November 1888. A new, heavier motor drove all four driving wheels through a single reduction geared motor. The center third rail for power collection was changed to a copper rod beside the rail, with a third rail "shoe" to draw power. Trial trips made over the next couple of months showed the *Franklin* capable of pulling as many as eight empty cars over the elevated. With a standard four-car train the locomotive was capable of pulling a train at up to 24 miles per hour. Early the following year the locomotive was even put into regular service with the regular steam trains. But despite the *Franklin*'s success in its final iteration, the Manhattan Railway managers decided to continue steam operation and the electric locomotive was removed from the line later in 1889.

In addition to Daft, tests on the New York elevated were also planned by the Electric Railway Company of the United States, which had been formed in 1883 to join the electric railway interests of Thomas Edison and Stephen Field. The Edison-Field men had contracted with the Manhattan Railway to operate a test car over the Second Avenue elevated line and had planned a test installation. Developed jointly by Edison, Field, and Charles Batchelor, who was in charge of the Edison shops, the plans incorporated one element that would later become the standard for electric railway power. Instead of using a separate electric locomotive, Edison-Field proposed simply to put electric motors on the trucks of each of the passenger cars, providing much better adhesion by having every axle powered, providing each car with its own motor, making each car, in effect, an independent operating unit. Just how these independent units would be controlled was not made clear.

Construction of the third rail for the Edison-Field project got started in May 1885, and the Pullman Palace Car Company of Chicago was contracted to provide a test car similar to those used by the Manhattan Railway. As it was described in electrical trade journals, the car was equipped with a 600-volt Siemens type compound-wound motor, one on each truck, with power transmitted to the motors and axles by pulleys and a leather belt. Electric power would be supplied from a power station with five Edison dynamos located

Electrical inventor Leo Daft operated test runs on New York's Ninth Avenue elevated line between 1885 and 1888, pulling standard elevated cars with the electric locomotive *Benjamin Franklin*.　*Library of Congress.*

in the Durant Sugar refinery on East 24th Street. An October 17, 1885, issue of *Electrical Review* described the new test car, still at the Edison works, that would be put on the tracks by New Year's Day. But if tests of the car were ever made, they remain unknown, and the trade journals had no further mention of the Edison-Field car. Apparently, Edison had lost interest in electric railways.

Several years later Stephen Field, now associated with electrical engineer Rudolph Eickemeyer, set out to test a new design for an elevated railway electric locomotive. This was a short "steeple cab" locomotive, with a center control cab and sloping compartments at each end. The locomotive was mounted on two four-wheel trucks, one of which was unpowered while the other was powered by an electric motor designed and built by Eickemeyer. The locomotive was tested on one track of the short 34th Street branch of the elevated. Electric power was supplied to a center third rail from a single Eickemeyer generator that could produce power at anywhere from 800 to 1,100 volts DC. The locomotive was assembled and put in test operation in September 1887, but the under-powered locomotive could only pull a single car at 8 miles per hour, which was of little interest to the elevated railway company, and the Field project was soon forgotten.

The Sprague Electric Railway & Motor Co. had not been included in the 1884–1885 plans for tests of electrification of elevated railways, but Frank Sprague was also among the group that was trying to develop a satisfactory electrification. Later, Sprague remembered,

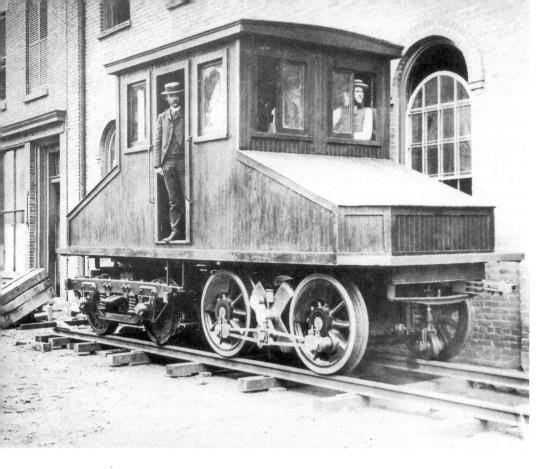

Electrical inventor Stephen D. Field built and tested this locomotive on New York's 34th Street El in September 1887.   *Industrial Photo Service.*

after a study of the movement of trains and the conditions of operation upon the Manhattan Elevated, I had schemed out a system, and in December of 1885 read a paper before the Society of Arts in Boston, advocating an equipment with motors under each car, and using shunt-wound machines to enable current to be returned to the line when decreasing from the higher to more moderate speeds.[17]

Sprague's December 10 presentation to the Boston Society conveyed a detailed study of the Manhattan Railway's Third Avenue line, which extended over about 8½ miles from South Ferry to the Harlem River, reaching a total of 26 stations on an undulating line that operated through a series of up and down or level sections. Operated with steam power, trains could attain an overall average operating speed of only 10 miles per hour. With 66 trains in operation during commission, or peak, hours, the Third Avenue line had very nearly reached its capacity. To increase capacity with longer trains would require the use of larger locomotives, which could not be accommodated on the existing elevated structure. Sprague's extensive calculations had been aided by locomotive expert Angus Sinclair; Col. Frank K. Hain, manager of the Manhattan Railway; and J. D. Campbell, general foreman of the Manhattan's machine shops.

While additional elevated lines or subways might be a long-range solution, said Sprague, there were two ways that the capacity of the existing lines could be increased. One would be to increase the number of cars in a train, if it could be done without increasing the weight of the locomotive. Another would be to increase the average operating speed, which would also require a larger locomotive. Neither of these seemed possible with steam power. "Hence," said Sprague, "we [are] obliged to turn to some other method of locomotion, and that which promises the most satisfactory solution is an electrical system. . . . I have for a long time been elaborating such a one," he said, "and am now convinced that this is the future method of propulsion for the trains of the elevated road, and it is a near future."[18]

Electric power in place of stream locomotives would permit the motors to be placed under cars and distributed throughout a train, providing an overall increase in power and allowing for longer trains without using a heavier locomotive. With power distributed throughout the train, the whole weight of a train would be effective for traction, allowing more rapid stops and starts, and increasing the overall average speed. The use of electric motors on the cars, too, Sprague noted, would permit more frequent service during off-peak hours. But how to control distributed power from a single controller had not yet been worked out, and it would be another decade before it was.

Frank Sprague had also developed the idea of power regeneration from a braking train or one descending a grade, and this would provide major cost-saving possibilities with the frequent starts and stops from an elevated train. Every motor coming to a station or running downgrade would become a power generator, feeding energy back into the third rail system. According to Sprague's calculation for the Third Avenue elevated, the maximum power output needed by electrification having regeneration capability would be reduced by more than a third from what would be required from electrification without it.

With the combination of reduced power demand from electrification, and with the much greater efficiency of developing power from a central plant instead of through locomotives, Sprague had estimated that he could cut the Third Avenue's coal costs by over 71 percent. Even further savings in maintenance of the elevated's overhead structures, Sprague assured his readers, could also be anticipated by eliminating the use of heavy steam locomotives.

In his closing argument about electricity for the New York elevated, Sprague also included a pitch for a Boston electric railway:

> With electric propulsion you can have rapid and smooth-running trains of one, two, or more car units. The strain on the structure being much less than in a steam plant, the whole structure can be made lighter in the same proportion.
>
> Dust, smoke, cinders, oil, and water will disappear.
>
> Power will cost less.
>
> Trains can be run at shorter intervals and under more perfect control.
>
> The energy of the train will become available for the purpose of braking.
>
> Repairs of the superstructure will be less.
>
> In short, electric propulsion, more than any other thing, will make practical for Boston what it has so long and so sadly needed—rapid transit to its suburbs. I need hardly point out to you the increase in the value of this property, which will more than pay the cost of the roads.[19]

Much of the next year was devoted to experiments on a section of elevated track in New York. Sprague first began operating a test car at the Durant Sugar Refinery on East 24th Street between late 1885 and 1886. A test car with a battery of five or six Edison generators, connected in series to get a 600-volt DC potential, and a section of track about 200 feet long laid between two adjacent buildings had previously been installed by the Electric Railway Co. of the United States. The Edison-Field tests then being inactive, Edward Johnson had arranged to lease the facility for the Sprague Electric Railway & Motor Co. Sprague had arranged a test car which was made up of the bottom framework of an elevated car equipped with two trucks and motors with single reduction gearing, shunt field control, regeneration, electric braking, and various operation and test equipment. Later on, a second pair of motors was installed with an interpole winding.

A key feature of Sprague's design for the test car was the adoption of a method for mounting the motors so that they were directly geared to the axle, solving what had been a source of endless problems for early electric railway designers. In what was sometimes called a "wheelbarrow" or "nose-suspended" mounting, one side of the motor was hung from the truck frame on a spring mounting, while the other was supported directly by the axle. Bearings in the axle side of the mounting permitted the motor to rotate directly about the axle, thus maintaining perfect alignment between the gearing on the motor shaft and the axle, no matter how irregular the track or the motion of the axle.

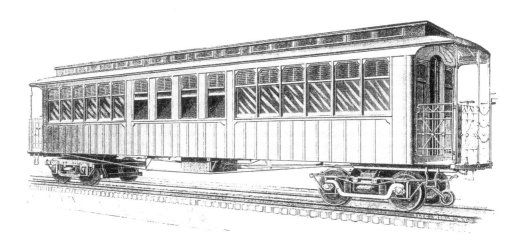

A drawing of a standard New York El car shows how Sprague equipment would be used for electric operation. One truck would be unpowered, while the other would be equipped by two electric motors. A center wheel would be used to collect power from a center third rail.   The Electrical World.

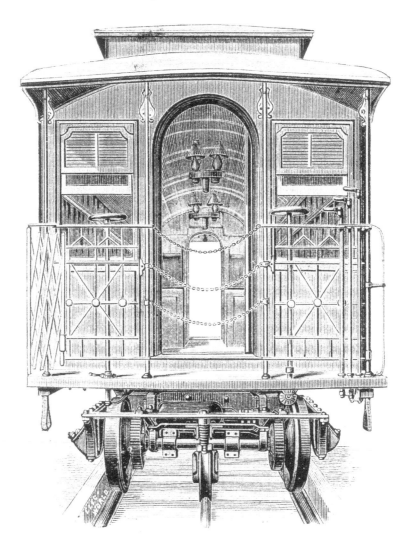

Three views of the Sprague car truck developed in 1885–1886 demonstrated the most important innovation by Sprague, which provided an effective link between traction motors and the wheels. This was the "wheelbarrow" or "nose-suspended" design for the traction motors.　The Electrical World.

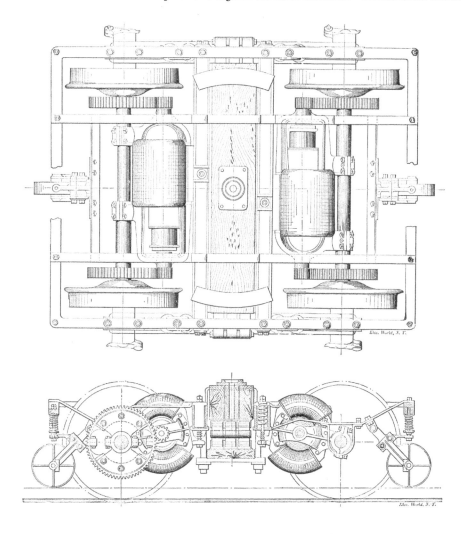

While earlier inventors had devised a similar arrangement, it was Sprague's application of this mounting that became an almost universal method for mounting electric motors.

Sprague conducted a series of tests over the next several months, often attended by the Field and Gould interests and others associated with the elevated railway. One such visit that didn't go so well, however, was a visit from Jay Gould and members of the Cyrus Field family—the principal owners of the elevated—and it has become an often-told story. During the operation of the test car, Gould stood near the controller and an open lead safety fuse. Sprague later recalled;

> Desiring to make an impressive demonstration of how readily the car could be controlled and braked in the short distance available for movement, I handled the controller rather abruptly, whereupon the fuse blew with a violent flash and Mr. Gould attempted to jump off the car. My explanation that this young volcano was only a safety device was not convincing and he did not return.[20]

By May of 1886 the experiments had been shifted to the elevated's 34th Street branch, and Sprague and his men were getting a new test car ready for service. The Manhattan Railway had made available a standard elevated car—No. 298—that would be outfitted for use as an experimental car. It was equipped with controls and a variety of other material required for electric operation, and two traction motors were mounted in "wheelbarrow" fashion on one of the car's two trucks. Sprague had decided on a 600-volt nominal voltage for the line, with the dynamos wound to supply an electromotive force of about 670 volts at their terminals, and spring-loaded wheels collected power from the center third rail. Among Sprague's men preparing the equipment were Fremont Wilson, Robert McPherson, James Brett, and Patrick F. O'Shaughnessy.

Even before the new test car was ready for service, Frank Sprague—who was vacationing in Richmond, Virginia, at the time—received a telegram from Edward Johnson on Friday, advising Sprague that he had promised Cyrus Field that he would show an electric car in operation on the coming Tuesday. Sprague and his men, of course, rushed to get the car ready:

In 1885 Frank Sprague set up this test car on the elevated railways of the Durant Sugar Refinery at East 24th Street. Sprague used the inactive power system that was set up by Thomas Edison and Stephen Field, and equipped the car with two trucks and motors, various electrical equipment, and operation and test equipment. *Frank Sprague Papers, Manuscripts and Archives Division, The New York Public Library, Astor, Lenox and Tilden Foundations.*

How we got together car body, truck and machines and completed a regulator, in face of a strike, and assembled them all by Monday night on the elevated railroad, is more than I can at present tell, but that night found me with my faithful assistants, McPherson and Crawford, by candle light making all the connections necessary to operate two machines at 600 volts potential by duplicate switches at each end of the car, and to brake the car as well, with no way of testing the apparatus or to ascertain the correctness of any connection. At 1 o'clock the next day there was an expectant crowd on the Thirty-Fourth Street platform, among whom were many men of prominence. While waiting for current half an hour passed with no evidence of "anything doing," save under my breath, and there Johnson was the object of many an earnest anathema for his strenuosity, for I did not know whether either machine would turn over, or whether, if they did so, they would operate alike—in short, everything was in the air. Finally, when current was put on the line I first tried one machine and then the other, but with no movement in response. Finally, in sheer desperation, I threw both machines into circuit, moved the regulator, and the car responded perfectly. For two hours every feat which could be tried with the machines was attempted without a failure.[21]

A short time before the 34th Street test was made, Superintendent C. E. Chinnock, of the Edison station in New York, offered Sprague $30,000 for a one-sixth interest in the Sprague Electric Railway & Motor Co. Even though he had, as Sprague put it, hardly enough money to pay his board bill, Sprague turned him down. Much impressed with the results of the just completed tests, Chinnock now offered Sprague $25,000 for a one-twelfth interest. Sprague accepted.

Encouraged by the results of the day's test, Sprague and Johnson were determined to continue the experiments. A second pair of traction motors was added during the summer, and metal resistances and water rheostats were tried. The experiments continued through December, but after the successful tests visited by Cyrus Field and other men from the Manhattan Railway, there seemed to be no further interest in the new power: "in all those months, so far as I can remember," said Sprague, "not a director or stockholder of the Manhattan road ever took the slightest interest in what was being done."[22]

Leo Daft and Steven Field did no better with their plans for electrification of the New York elevated railroads, and it would be another decade and a half before the Manhattan roads would finally begin the conversion to electricity, well behind the elevated roads in Chicago and Brooklyn.

These drawings depict a typical series-wound direct current electric motor. Principal components include the rotating electromagnet armature, made up of an iron core on which are wound coils of copper wire; the commutator, which is mounted at the end of the armature shaft, and which transmits the current through the circuits of the armature; the brushes, which are usually carbon brushes that connect the armature with the field-magnet coils and external circuit through the commutator; and the field-magnets, which are stationary electromagnets surrounding the armature. These are usually bipolar motors, fitted with insulated copper wire wound in a manner to produce opposite polarity in the opposite pole-pieces. The first figure shows a very schematic arrangement of a series-wound electric motor, with electric power passing from the generator through the brushes to the commutator, through the rotating armature, and back through the brushes to the field magnets. The second shows a drawing of a 27-horsepower General Electric 52 motor, with the armature inside the motor housing, which contains the field-magnets. The third is a cutaway drawing of the G.E. 52 motor armature. *Middleton Collection.*

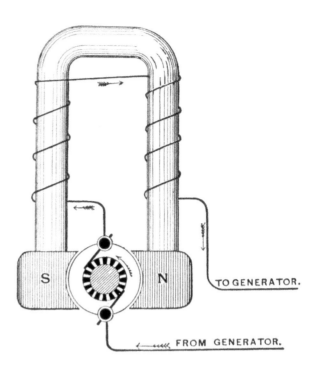

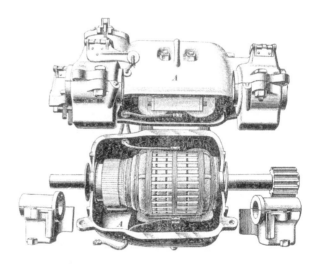

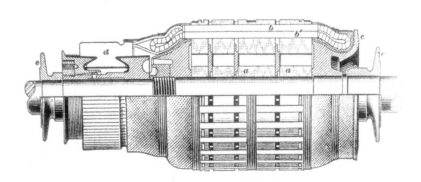

Though Frank Sprague failed to gain the hoped-for work on the New York elevated, his work there from 1885 to 1886, more than any other, set in place the almost uniform practice that would later be followed for elevated and subway rapid transit electrification.

> The machines used on these experiments may be termed the parent models of the modern electric railway motor. They were centered through brackets on the driving axles and suspended at the free end by springs from the transom, the elliptic springs being interposed between the support and the car body. The motors were single-geared to the axles, had one set of tilting brushes, were run open, and they were used not only for propelling the car but for braking it. At first shunt wound, increase of speed being first accomplished by cutting out armature resistance and then inserting the resistance in the field, there was added later a correcting coil in series with the armature at right angles to the normal field to prevent shifting at the neutral point.[23]

Sprague had moved rapid transit electrification from the practice of trains of cars pulled by locomotives to the concept of self-propelled, decentralized "multiple unit" electric cars. The only important remaining element was the development of an effective way to handle the centralized control of decentralized motors. In an 1886 paper, Sprague wrote about doing this in some vague kind of a regulating truck: "If the motors are thus placed under the cars, each can be made an independent unit, or a dozen cars can be operated in a single train by a small regulating truck placed ahead of them."[23] A decade later, Sprague would come up with the best answer to this, too, the "multiple unit control."

In the meantime, with the New York elevated railroads still unwilling to invest in electrification, Frank Sprague was soon off to electrify the street railway.

4

TRIUMPH AT RICHMOND

New York City began operating the first American animal-drawn street railway in 1832, and over the next several decades the horsecars had spread to other cities as the country's urban population grew. By the beginning of the 1880s urban public transportation had grown into an enormous industry, and one that was steadily growing ever larger. In 1881, for example, there were some 415 street railway companies in the United States, and they operated 18,000 cars over 3,000 miles of line and transported well over one billion Americans every year.

But animal railways left much to be desired. They were slow, averaging only about 4 or 5 miles an hour, and the capacity of a car was limited. They were expensive as well. The work was extremely demanding for the animals, and it took great numbers of horses and mules to power them, usually requiring about eight to ten animals that had to be fed to keep each car in service. They could last only a few years in the arduous service, and thousands of new horses had to be acquired every year.[1]

Recognizing that the horse was what historian John H. White, Jr., called "the weak link in its operation chain," there were early attempts to find a more

Artist Jay Hambidge pictured the new streetcars in front of a Richmond theater for an article by Frank Sprague in the August 1905 issue of *The Century Magazine,* recalling his great success of 1887–1888.

satisfactory form of mechanical power. One of the first was the use of a steam locomotive to pull the cars. These were generally small-scale machines, often shrouded inside a "dummy," usually a wooden shroud that made the locomotive look more like a passenger car. This, it was hoped, would avoid upsetting passing horse-drawn vehicles. But the smoke, cinders, and noise of the locomotives made them unpopular in city streets, and steam operation was inherently inefficient in the stop-and-go service that the cars had to provide. Many variations were tried, but few ever thought that they ran well enough to replace horsepower. There were several attempts to develop some form of "fireless" engine, which used compressed air, ammonia gas under pressure, or a solution of caustic soda for heating the water in the boiler. But these ideas, too, failed to attract much interest.[2]

Of all the inventions and ideas brought forward to mechanize the street railway, only one achieved much success in the years before the electric street railway became practical. This was the cable railway developed by San Francisco businessman Andrew S. Hallidie, who proposed the idea of using the tough steel cable that he was already employing for such applications as elevators and long aerial tramways. Hallidie conceived the idea of using an endless, continually moving wire rope that would be pulled through a trough between the rails and below the street, and would be powered by a central steam plant, whose almost limitless power could surmount the steepest of hills and carry the heaviest loads. A special contrivance called the grip would permit the operator of the car to firmly grip the cable or release it to permit the car to be stopped at any point.

Hallidie's first cable line opened in San Francisco in August 1873 and quickly proved a success, and the next two decades saw the rapid development of the new cable railways. San Francisco's hilly streets were well suited for the cables, and by the late 1880s the city had more than 100 miles of line operating over eight different routes. By 1894, when the cable railways reached their peak, there was a total of some 360 miles operating in 29 cities in the United States, and still others were operating as far afield as New Zealand and Great Britain. American cable railways operated nearly 5,000 cars and carried 400 million passengers a year.

The cable cars were a marked improvement over the horse railways. Operating at about 9 or 10 miles per hour, they were almost twice as fast as the horsecars, and they could carry much larger loads. But the cables had some serious shortcomings as well. Because of the complicated underground trough and all of the cable railway installation that were required, they were extremely costly to build—anywhere from $100,000 to as much as $300,000 per mile, enormous numbers at the time. Because of the great weight of the cables and their associated equipment, a disproportionate share of the energy delivered by the steam plant was absorbed just by the moving parts of the system; on some lines this absorbed as much as 75 percent of the power plant output. One of the longest cable runs was made by New York's Third Avenue Railway, which was 13 miles in length and weighed an estimated 260 tons.[3]

Even as new cable railways were still under construction, there was steadily growing interest in applying the new wonder of electric power to the problem of powering a street railway system. As early as 1883, when the cable car boom had scarcely begun, H. H. Littell, the president of the American Street Railway Association, proclaimed in an address in Chicago: "I see in the recent subjugation of the subtle and hitherto elusive force of electricity to the needs of man boundless possibilities for the world's three great requisites of advancement: heat, light, and motion."[4] Indeed, almost a half century before Littell spoke inventors had begun the experimentation that would eventually lead to the successful development of electric traction.

In 1835 Thomas Davenport, a blacksmith and inventor from Brandon, Vermont, built and operated a small circular electric railway that was one of the earliest attempts at electric transportation. Just a few years later, in 1838, Robert Davidson of Aberdeen, Scotland, built a 7-ton electric locomotive that ran over the Edinburgh-Glasgow Railway. Professor Moses G. Farmer of Dover, New Hampshire, in 1847 demonstrated a small electric locomotive that could carry two people, and three years later Professor Charles G. Page of the Smithsonian Institution operated an electric locomotive between Washington, D.C., and nearby Bladensburg, Maryland. All of these early experiments drew their electric power from a wet cell battery, and if they demonstrated anything it was that batteries were an unsatisfactory and uneconomical source of power for an electric railway.

An important advance in electric railways came in 1879, when the German electrician and inventor Ernst Werner von Siemens successfully demonstrated a small electric railway at the Berlin Industrial Exhibition that was powered from a central dynamo, or generator, power supply. Two years later, Siemens installed a mile-and-a-half-long electric railway in Lichterfelde that was the world's first regular fare-paying electric railway. The generator finally provided a workable power supply for the electric railway, and there were soon several

Thomas Davenport's early attempt at an electric railway was operated by an electromagnetic motor that was powered by a battery. The equipment is now on display at the Smithsonian Institution. *Smithsonian Institution, National Museum of American History.*

inventors hard at work trying to solve other problems of the electric railway. Siemens went on to head the enormous German electrical company that remains today a major international firm.

Two other inventors developed dynamo-powered electric railways soon after Siemens' experimental operation in Berlin. During 1880–1881 Stephen D. Field, a son of transatlantic cable promoter Cyrus Field, built and operated an experimental electric locomotive in Stockbridge, Massachusetts, while in the spring of 1880 inventor Thomas Edison developed and operated a short loop of track at his Menlo Park, New Jersey, laboratories. The small electric railway was powered by a dynamo, or generator. Edison developed an improved locomotive the following year, and in 1883 Edison's interests were joined with those of Stephen Field to form the Electric Railway Co. of the United States. Edison, then much occupied in development of electric illumination, took little part in the work of the new company, which was managed by Field. A new experimental 3-ton locomotive, *The Judge,* was completed in June 1883 and exhibited on a track laid in the gallery of the main building at the 1883 Exposition of Railway Appliances in Chicago, but none of them were ever further developed.

Most of these early electric railways were developed as experimental or exhibition lines, but several new electric railway builders were now beginning to look at commercial transportation. One of the most important of these was Leo Daft, a British-born inventor who in 1879 had started developing electric light systems and dynamos, and by 1882 had begun developing motors for electric traction. The following year he constructed a 2-ton electric locomotive called the *Ampère* that successfully carried 75 people up a sharp grade in upstate New York. After operating several other exhibition lines, including some tests of electric power on the New York elevated system, Daft was engaged

A year after Edison's installation of the small electric railway in Menlo Park, New Jersey, this second, larger electric line was built, in which the electric locomotive was designed to look like a steam locomotive. *Middleton Collection.*

in 1885 by Thomas C. Robbins, the general manager of the Baltimore Union Passenger Railway, to electrify the railroad's 2-mile line to Hampden. Daft installed a powerhouse, a third-rail power distribution system, and two little four-wheel locomotives—Daft called them "tractors"—which could each pull up to two standard horsecars. The line was ready for operation in August 1885, the first U.S. electric line operated by electric power for an extended period, but after more than four years of operation the line was reverted to horsecar operation in 1889. The line was popular with its passengers, but the owner was unwilling to make the needed improvements.

Over the next several years Daft built several other electric lines. In April 1887 a Daft system was opened on Pico Street in Los Angeles, California. Daft had switched to a double overhead wire power system, but the line's several four-wheel locomotives were almost identical to those used in Baltimore. The line continued to run until June 1888, when a boiler explosion at the power plant and bankruptcy ended the service. By far the most successful of the Daft electrifications was opened in September 1887, when the Sea Shore Electric Railway in Asbury Park, New Jersey, began operating a 4-mile line with eight

The Baltimore Union Passenger Railway operated a 2-mile electric line developed by Leo Daft between 1885 and 1889 before it reverted to horsecar operation. *From the February 13, 1886, issue of* The Electrical World.

cars in a service which continued in operation until 1931. Like the Pico Street line, this one, too, adopted a two-wire overhead power system. Daft went on to electrify more than a half dozen other electric lines before finally selling his electric railway patents and going on to other enterprises.

John C. Henry, a Canadian-born former farmer and later a Kansas City telegraph operator, began to experiment with a novel car powered by a manual transmission for a constant-speed electric motor which drew its power from an overhead "trolley" or "carrier" that ran over two parallel electric wires. The first Henry experimental car was tried early in 1885, but was soon taken out of service. A second car, using the same basic technology, operated with more success from 1885 to 1886 over a short section of the Kansas City, Fort Scott & Memphis Railroad, and later, in 1886, Henry installed four electric cars along a mile of Kansas City's East Fifth Street for regular passenger service. The line continued to run through the end of the year but without much success and was shut down. Henry got one more chance to install his electric cars when he

After developing an electric railway system operating from an underground power supply, Edward M. Bentley and Walter H. Knight opened a 4-mile line in Allegheny City, Pennsylvania, in January 1888, using both an overhead and underground power supply.   *Library of Congress.*

was contracted to install a new line in San Diego, California. Using the same technology, the Electric Rapid Transit Street Railroad Company opened over a 5-mile line late in 1887. This time the line continued to operate for about a year and a half, before operating losses brought it to an end, to be replaced by a cable railway which did no better.

Two Cleveland electric railway inventors, Edward M. Bentley and Walter H. Knight, tried a different solution to the difficult problem of current collection. Bentley was a patent attorney, while Knight had a background in mechanical engineering, and the two joined forces in electric railway development. Recognizing the problems that other inventors were experiencing with overhead power systems, as well as the likelihood of growing opposition to unsightly overhead lines, Bentley-Knight electrified a 2-mile section of the East Cleveland Horse Railway Company in 1884 using an underground power conductor placed in a slotted wooden conduit between the rails, with a "plow" placed on the car being utilized for current collection. The line opened in July

1884 and continued to operate for a year and a half. A second Bentley-Knight line, opened in Allegheny City, Pennsylvania, in January 1888, used both the underground power system in the downtown area and a two-wire overhead system elsewhere. The line ran well, but was converted two years later to the Sprague system for uniformity with other systems in the Pittsburgh area.

Still another electric traction inventor was Sidney H. Short, a professor of physics at the University of Denver. Like Bentley-Knight, Short also favored an underground conduit system, devising an unusual "series" system of current collection in which the power feed was installed in a conduit between the two rails. A series of contactors, or "circuit breakers," was installed in the conduit at a spacing of 17 feet, while on arms projecting to the front and rear of each car were metal shoes which slid along the slot. Suspended on hangers from the shoes was a long "brush," a hickory slat on which were mounted brass strips wired to the motor. The brass strips came in contact with the contactors to complete a series circuit through the motor. A switching coil permitted the speed of the motor to be varied, or the circuit to be bypassed for a stopped car.

Short conducted experiments in 1885, using a small electric car operated over a loop of track in a university building, and the following year he began running a 3½-mile line of the Denver Electric & Cable Railway Company on Fifteenth Street. The line continued to operate for the next few years, but the complexity of the electric system and dirt and water problems in the conduit made it less than satisfactory, and a cable railway replaced the electric line. Professor Short went to Ohio in 1889 and built several lines using some variations of the design used in Denver. His patents were later sold to another company, and eventually landed at Westinghouse.

Foremost among the group of inventors who had brought the electric street railway to the edge of commercial success in the late 1880s was Belgianborn Charles J. Van Depoele. Immigrating to the United States in 1868, Van Depoele established a successful church furniture-manufacturing company in Detroit, while continuing his electrical studies and experiments. By 1881 Van Depoele was illuminating the streets of Chicago using a dynamo of his own design, and he soon became interested in electric traction. During the winter of 1882–1883 he built an experimental car powered from an overhead wire, and the following fall he built a second car which was operated at the Industrial Exposition in Chicago. In 1884 Van Depoele took on a much more ambitious exhibition at Toronto's Industrial Exposition, operating a train powered by an underground conduit, and capable of transporting as many as 200 passengers at a speed of 30 miles per hour. The following summer Van Depoele returned to the Toronto exposition with a bigger and better train and a locomotive that could pull three cars. Operating all day, the train could transport more than 10,000 people in a day. A major technical advance was Van Depoele's development of an overhead power wire. Current was drawn from the wire by a contact wheel, carried on a pivoted beam (later, usually called a "trolley pole"), that was held against the underside of the wire by spring tension.

The 1885 Toronto exposition was a huge success, and Van Depoele was quickly deluged with orders for the installation of new electric lines. A new Van Depoele exhibition line opened at the New Orleans Exposition only a month after the end of the Toronto show. And before the end of the year Van Depoele had completed work on a 1¼-mile electric line in South Bend, Indiana, which included four cars. In a major departure from his use of an overhead under-running trolley pole, he had returned to a system, much like those used by Leo Daft or John Henry, with power collection provided by a traveler or trolley that ran over the overhead system, with a flexible cable used to carry power to the car. By mid-1887 Van Depoele had more than a dozen new electric lines either under construction or already operating. Most of these were small installations, but a project in Montgomery, Alabama, was by far the largest yet built.

Construction of the Capital City Street Railway Company got underway in Montgomery early in 1886, and two cars were in operation by April. Electrification was extended and by June 1887 the line operated 18 cars over 15 miles of track, making Montgomery the first completely electrified street railway system anywhere in the United States. Montgomery found that the overhead system linked with travelers or trolleys was unsatisfactory, and Van Depoele soon changed the line to use an under-running trolley pole like that used in Toronto. The line was at least successful enough to gain a permanent status, interrupted only for a temporary return to mule power after an 1888 fire destroyed the power plant.[5]

At the beginning of 1888 there were only 21 electric railway companies operating, with some 172 electric cars operating over a grand total of 86 miles of track. This was still a very small start, representing scarcely 1 percent of total U.S. street railway mileage, but electric traction offered enormous possibilities as the technology began to achieve the reliability and performance needed for commercial success.

There were still some difficult problems to be solved. Among the most important of these included the means of transmitting electric power from

TRESSLAR.

MONTGOMERY, ALA.

VAN DEPOELE ELECTRIC RAILWAY as
in practical operation on the lines of the
Capital City Street Railway, Montgomery, Alabama.
Grades on this road varying from four
to seven and one-half per cent.

Van Depoele Electric Manufacturing Co.

Foremost among Sprague's competitors was Charles J. Van Depoele, who by mid-1887 had begun operation of this line in Montgomery, Alabama. Not visible here is the two-wire overhead power system that Van Depoele used. *The Smithsonian Institution, National Museum of American History.*

the stationary plant to the electric car, the mechanical connection between traction motors and the powered wheels, and a variety of problems with the traction motors themselves.

Power supply to the electric car ranged from power supplied through the use of the two running rails, using one rail as the positive and one as negative; third-rail positive rails and underground conduits; and a variety of overhead power wires. Edison and Field used the two running rails, usually with a positive center third rail, while Daft used both the running rails and overhead wires on his various systems. Power was supplied at anywhere from 100 volts DC, to as much as 280 volts DC, but even these low voltages were problematic. In his Baltimore installation, Daft found that he had to interrupt the third rail at crossings for safety, substituting an overhead gas pipe to provide power across the road. The underground conduits developed by Bentley-Knight and

Sidney Short were both costly to build and often troublesome to maintain, and these underground systems were usually used only when required by municipal authorities. Overhead wiring proved to be almost the universal choice for street railway electrifications. The two-wire system was used on some early systems but almost all were changed to single-wire, using the overhead wire as positive and the running rails as negative. The only major exception to this was in Cincinnati, where a two-wire system was always maintained.

An early problem with overhead power collection was a reliable means of making continuous contact between the wires and the moving car. Early two-wire overheads used some form of four-wheel cart, usually called a "traveler" or "trolley," that ran above the wires, drawing power through the wheels. Power was transmitted from the traveler to the car through a flexible connection, usually connected to the roof of the car. A similar traveler was used on a single-wire system. All of these travelers were troublesome. Sometimes the traveler would fall off, crashing down on top of the car, and they were always difficult to handle when cars had to pass or run through switches. The under wire connection developed by Van Depoele in 1885 apparently worked well, although he returned to some type of traveler current collection in subsequent projects.

Finding an effective connection between the traction motor and the wheels of the electric car remained a continuing problem. Edison's 1880 venture used a friction motor to make the connection, and later substituted a belt drive for the connection. Bentley-Knight later also tried a friction drive. Daft and Van Depoele both used some form of belting to make the connection, and Daft later tried a geared connection while Van Depoele also used a chain drive. Bentley-Knight tried a steel wire cable for the connection. But all of them were troublesome.

Frank Sprague himself, working on designs for proposed New York rapid transit cars from 1885 to 1886, had come up with what proved by far to be the best solution, an arrangement sometimes called a "wheelbarrow" mounting that geared one end of the traction motor directly to the axle, while the other was hung from the truck frame on a spring mounting.

Traction motors themselves often experienced problems. Among the most troublesome on early motors were the brushes that transmitted power to the armature. Most early designs used brass or copper brushes, which wore out quickly and required frequent maintenance and adjustments.

Up until 1887, Frank Sprague's electric railway interests had been directed to rapid transit. In the mid-1880s this was confined only to New York City, and despite two years of his presentations on the benefits of electric power and all of his tests and demonstrations, the financiers who controlled the New York elevated railways—Jay Gould and others—were uninterested in making the substantial investment that would be required for electrification. They were already making plenty of money.

Frustrated by the lack of opportunity he was finding in rapid transit electrification, Sprague had decided to shift his attention to street railways. With the number of electric railway inventors already at work on street railway development, it would hardly have seemed that another one was needed. But Sprague would prove to have the strong technical ability and experience needed to solve the problems still preventing large-scale electric traction from achieving reliability and profitability, and he would have the determination and commitment that would be required to get it done. Beginning early in 1887, Sprague successfully contracted for small street railway electrifications in Carbondale and Wilkes-Barre, Pennsylvania; Wilmington, Delaware; and St. Joseph, Missouri. But far more important to Sprague was the opportunity that came to him for electrifying a new street railway in Richmond, Virginia. It would be by far the largest electric railway ever undertaken anywhere in the world.

Richmond's plans for a new street railway system had, at least in part, a political background. During the post–Civil War period the city had grown and added three new "suburban" wards. The existing horsecar company had not been willing to expand into these new areas, and these outlying wards had long suffered from an inadequate public transit system. The local branch of what was then called the Readjuster Republican Party saw this as an opportunity to strengthen its progress and reform efforts against the more conservative Democrats. John S. Wise, who had been a leader in the party and a strong supporter of better transportation to the outlying wards, began planning for the new street railway in 1886. The initial 12-mile system was designed to serve both the older parts of the city, as well as to provide service to the newer outlying wards, which happened to be largely Republican.

The Richmond City Council approved construction and operation of the new Richmond Union Passenger Railway in March 1887, despite the opposition of the existing horsecar line, which belatedly had proposed to expand its own system. With a slate of Reform Republican candidates (as they were now called) running for November 1887 election, the reformist group wanted to have the new cars running into the outlying wards before the election, figuring that this would help them win.[6]

Soon after the construction of the line was approved, the city council also approved a proposal by the new railway to operate with electric power. New

York financier Maurice B. Flynn, who helped John Wise secure outside financing for the project, was familiar with Frank Sprague's recent work in proposing electrification of the New York elevated railways, and brought Sprague into the Richmond project for the electrical work.

The proposed contract for Sprague was a daunting one. The plan for the 12½-mile line called for the completion of the electrical work in only 90 days, with the track not yet built and even the exact route sometimes not yet established. Sprague would have to provide a steam and electrical plant of 375-horsepower capacity and a complete overhead system, and he would have to equip 40 cars with a total of 80 traction motors and all their appurtenances. Each of the four-wheel cars was to be equipped with two of the 7½-horsepower traction motors, mounted in the "wheelbarrow" arrangement that Sprague had developed several years earlier for experimental cars on the New York elevated. Thirty of the cars were to be operated at one time, with grades as high as 8 percent to be ascended. The price for equipment was to be $110,000, to be paid in cash after 60 days of satisfactory operation. At the time of taking the contract, Sprague had only a blueprint of a railroad machine and some experimental motors used on the elevated road, and 101 details essential to success were still undetermined.

"Considering all of the difficulties which confronted the company," remarked Sprague several years later, "it is probable that, had they been known beforehand and the solution of them unforeseen this contract would never have been made."[7]

While Sprague may not have known of all the problems to be confronted, there is little question that he knew it would be difficult. But he knew too that the successful completion of this unprecedented project would stand as the prototype for a modern electric railway system, and that it would have enormous benefits for its engineer. Despite his youth—he was then only 30 years old—Frank Sprague was supremely confident of what he could do, and approached any difficult problem with an unflagging energy and commitment until it was successfully completed.

Sprague was joined in the Richmond project, too, by a remarkably capable group of talented electricians and engineers who wanted to be part of the dramatic new work in electric transportation. A number of them came from the military services, whose education at West Point and Annapolis was among the best in science and mathematics that could be found anywhere at the time.

Chief among them were S. Dana Greene, who would head up the installation work in Richmond, and Oscar T. Crosby, who would oversee work at the firm's New York factory where the electrical equipment was being manu-factured. Greene had graduated from Annapolis in 1883 at the top of his class, just four years behind Sprague, and joined Sprague in 1887. Crosby was an 1882 graduate from the U.S. Military Academy who resigned his commission in 1887 to join Sprague. Years later, Crosby remembered the experience of joining Sprague: "Can you get a week's leave and come to New York at my expense?" Sprague had asked Crosby.

> So, I came, I saw, and I was conquered. Resignation of my commission in the Army followed. Thereupon, for several years, I drew no sober breath. We were all intoxicated—drunk with the work inspired by your indomitable energy, your penetrating intelligence, your bold pioneering through obscure problems. Out of the woods you emerged, flags flying, as we captured Rich-mond, Wilmington, Scranton, St. Joe. Great days! Technical difficulties, financial difficulties, professional jealousies, nothing could daunt the man to whom the world owes more, in the matter of electric traction, than any other ten men combined.[8]

Altogether, there were more than a dozen electricians and engineers that formed this "engineering corps at Richmond, Va., 1887," as *The Street Rail-way Gazette* called it. Ayers Derby Lundy was an electrical engineer who had graduated from both Princeton and Cornell universities, and who would be Sprague's chief engineer. Horace F. Parshall was Sprague's assistant on the design of the traction motors. Other key men included William Bates, G. H. Preble, F. Wyse, D. Lewis, J. Picket, G. T. Baker, H. P. Niles, J. Marshall Atkin-son, Frank Lewis, David Mason, and W. H. Forbes.

G. W. Mansfield was the only experienced motor man of the entire corps, having previously connected with traction pioneer Leo Daft. "He brought with him to this hopeful band that priceless jewel—experience—which, when lacking in an experience of this kind, is absolutely ruinous," commented *The Street Railway Gazette*.[9] Pat O'Shaughnessy was a resourceful mechanical man who had worked with Sprague since the New York elevated tests, and who, as Sprague put it, "is possessed of a most happy mechanical judgment."[10] But O'Shaughnessy was more than just a good mechanical man, for he was cited by Sprague as a co-patentee for no fewer than four patents developed on the Richmond project.

Sprague was notoriously short tempered when things went wrong, and O'Shaughnessy was often the butt of his outbursts. "One of the often re-peated stories was the frequency with which you told [O'Shaughnessy] 'to go to hell,' and his usual reply that 'he did not care to visit your home when you were away,'" remembered street railway man S. W. Huff about the Richmond project.[11]

The project began to have setbacks and run into problems almost immediately. The contract had been awarded to the Sprague Electric Railway & Motor Co. in May. In July Sprague was laid up with an attack of typhoid fever, which kept him away from the work over a nine-week convalescence and forced him to leave the work to his assistants. Finally returning to work on October 1, Sprague looked over the condition of the work for the first time, and it was not reassuring.

"The track was simply execrable," said Sprague, "laid for profit, not for permanence, and with no adequate idea of the necessities. It was a flat 27-pound tram-rail of antiquated shape, poorly jointed, unevenly laid in red clay, and insecurely tied together." Instead of the 8 percent maximum grade previously discussed, the line along Franklin Street would reach grades of as much as 10 percent, and Sprague was uncertain of how his electrical equipment would handle it. "The sight of the main grade was one of my most unpleasant sensations. There were two things which seemed probable," he said. "The first was that the car, for lack of adhesion, would not ascend the grade at all no matter how powerful the machines, and the second was that I knew the latter were lacking in capacity."[12]

Sprague began thinking of a design for an electric motor–driven cable system to be installed in pits sunk beneath the track to help carry the cars up the steepest hills. But before deciding this, suggested Edward H. Johnson at a meeting in New York, let's "find out whether the car can get up the grade at all."[13]

Late on the evening of November 7, Sprague, Greene, and a crew, accompanied by streetcar superintendent George A. Burt and a reporter from the *Richmond Dispatch,* took a car out of the Church Hill car shed at 29th and D streets to try the experiment. At the foot of a steep hill the car got locked in a sharp curve. With difficulty, Sprague got the car pulled out of the curve. He was still doubtful about the hill ahead, but Burt said, "If you can get out of such a curve as that we just left, you can go up the side of a wall."[14]

With Sprague himself at the controls the car made its way up one hill and then another, easily swung through a sharp curve on a 6 percent grade, and finally climbed steadily to the top of the long Franklin Street hill, where it came to rest amid an enthusiastic after-theater crowd. By this time the car's overburdened motors were boiling hot, and there was a peculiar bucking sound

This 1888 photograph shows both the steepness—up to 10 percent—on either end of the Franklin Street hill through downtown Richmond, and the poor quality of the track structure that Sprague had to work with. *From the August 1905 issue of* The Century Magazine.

that told Sprague that there was a short-circuited armature, a difficulty that would become all too common in the weeks ahead. While Sprague waited for the crowd to disperse, Greene set out to get some "instruments," which turned out to be four sturdy mules. The car set off to return to the car shed with the defective motor cut out, while the mules followed at the rear. But on the curve at Bank and Twelfth streets the car derailed, crashing into the base of the St. James Hotel with significant damage to the car. The mules brought it the rest of the way home.

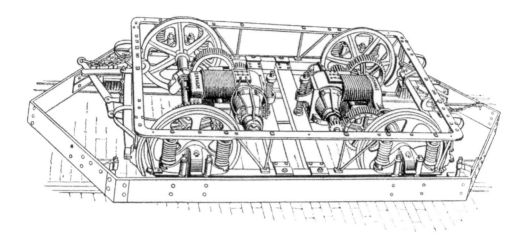

The Richmond motor truck as it was modified in 1888, showing the installation of double reduction gearing to permit the cars to operate over the line's 10 percent grades with their small 7½-horsepower motors.   *From the August 1905 issue of* The Century Magazine.

Nevertheless, as reported in the next day's issue, the *Dispatch* man was much impressed with the trial:

> It is a success!
> It is a revolution!
> It travels over more than two miles of track in Richmond!

Having at least proved that the traction motors could ascend a 10 percent grade, Sprague went back to work in New York to provide a solution to the problem of the overheated traction motors. This could be done by modifying the car's motors from a single reduction arrangement to a double reduction gearing, which required a whole new set of gears and mountings. Sprague designed the new arrangement and persuaded the Brown & Sharpe Manufacturing Company in Providence, Rhode Island, to complete the work on a rush basis without regard to the expense.

These problems solved, the cars began to venture out on the line more frequently, but new problems kept appearing. The patented "wheelbarrow" mounting of the traction motors developed by Sprague worked well, but other elements of the traction motors required frequent modifications as the severe strains of operation over the line's rough track revealed one weakness after another. Field-magnets, for instance, hastily wound in sections with crude insulation, frequently grounded and short-circuited.

The overhead construction was most unsatisfactory. Sprague had decided on using an under-running trolley for current collection, similar to that tried by Charles Van Depoele at the Toronto exposition in 1885, but getting it to work well proved to be extremely difficult. Sprague had trouble acquiring a suitable trolley wire until the Aluminum Brass & Bronze Company of Bridgeport, Connecticut, produced an alloy wire that met all the requirements for conductivity and tensile strength. Switches in the overhead work were causing trouble. No less than 50 to 60 under-running trolleys of almost every conceivable shape and character were tried until Eugene Pommer, one of Sprague's draftsmen, finally found one with a swiveling trunnion (something which can be rotated or tilted), turning in a tripod and carrying a trolley pole with a retrieving line.

Sprague's men had been under intense strain, with day and night work the rule. Costs were out of control as unforeseen problems were dealt with. The company's promoters pressed Sprague to begin service, but still not satisfied with his installation when the 90-day limit had been reached, he had to agree to a reduced payment of $90,000 in order to obtain a time extension. "The road must be made to go"; remembered Sprague, "it meant financial ruin for many unless it did. To the railway development its failure would prove a most serious blow. My own reputation and future career, as well as that of my associates, seemed blasted if failure marked the Richmond road."[16]

By January 7, 1888, the company was able to operate nine cars throughout the day and transported several thousand delighted Richmonders. Two days later the company attempted to begin revenue operation with six cars, and Motorman P. N. Grant and Conductor Walter Eubank took the first car out of the 29th and P streets car shed, and a Church Hill resident, William A. Boswell, presented the first revenue nickel to Conductor Eubank.[17] The cars operated only sporadically during the day, and they were soon on their way back to the car shed for still more modifications.

During the next several weeks a few cars continued to operate intermittently, and by the end of January the company was ready to try regular service again. As a preliminary run, the cars spent a day carrying children of the town, and on February 8 the line opened for regular service. The cars operated without difficulty the first day, but the next day they began to lock, and car after car suddenly stopped dead in their tracks. Sprague was convinced that the castings were faulty or the gears improperly cut, but Pat O'Shaughnessy insisted that the problem was simply a lack of adequate lubrication. More oil was applied, and the cars were soon running again.[18]

Demonstrating his versatility, Sprague mechanic Pat O'Shaughnessy climbed aboard the roof of a Richmond streetcar to clear the trolley wire of sleet and keep the line running. *Artist Jay Hambidge pictured the incident for the August 1905 issue of* The Century Magazine.

On one cold winter morning Sprague awoke to find the trolley wire covered with sleet and icicles, and then soon saw Pat O'Shaughnessy riding on the roof of the day's first car, wielding a broom to clear the line.

Motor problems continued to plague the line. The brushes continued to be one of the most persistent trouble sources:

> At first [we tried] flat ones, solid and laminated, which would wear through, double over and hug an arc of the commutator, or oblique, solid, and laminated brushes, which would catch in the bars, reverse or split and straddle half a dozen bars, with the result of ruined commutators, crossed armatures and burnt-out fields. Then we tried tilting-brushes, and various shapes of copper, bronze and brass, set on end and pressed down by springs, and also solid bars of brass.[19]

But none seemed to work.

"The track soon looked like a golden path," Sprague later recalled, "for the rough commutator bars acted like a milling cutter, and shearing off the ends of the contact bars would send a shower of shimmering scales over machines and roadway."

At this point, Sprague was using about $9 worth of brass daily just for brushes, and a car was unable to complete even half a trip without a stop for inspection and generally a change of brushes. Charles Van Depoele, who had had the same problem with his motors, suggested the solution, which was the use of carbon brushes.[20]

Armatures were disabled so regularly that replacements often had to be shipped from New York by express in order to keep the car running. At times equipment was maintained in service only by borrowing parts from other cars. "Greene, this is hell," Sprague had commented in a message at about this point.

But eventually the problems lessened, and the cars kept running. Gradually the number of cars in service was increased from 10 to 20; by May 4 the service had been increased to 30, and for the first time the company was able to provide service over its entire system. Soon after Sprague could operate 40 cars at one time, 10 more than he had contracted for. Operating costs were only 40 percent of what they would have been with horses. On May 15, 1888, the Richmond Union Passenger Railway Company informed the Sprague company that the contract had successfully filled all the terms and conditions of the contract, and accepted the work.[21]

"It is almost needless to say that on that day we felt that we owned the city and the street as well," Sprague later recalled. "Fatigues and worry were all forgotten in what was to us a supreme moment."[22]

But there was far more in hand than just the successful completion of the project in Richmond. The magnitude of Richmond's success would quickly spread, and it would set in motion an extraordinary shift in street railways to electric power. In New Orleans, Louisiana, a circular demanding a mass meeting calling for electrification came up with the now-famous sign: "Lincoln set the negroes free! Sprague has set the mule free! The long eared mule no more shall adorn our streets."[23]

Boston's West End Street Railway Company, then the world's largest street railway system with some 212 miles of track, 1,700 cars, and 8,000 horses, was contemplating a change in motive power. The West End's general manager, Daniel F. Longstreet, was a firm advocate of the cable railway, but West End president Henry M. Whitney, together with Longstreet and his board of directors, was persuaded to visit the Richmond line, also visiting the new Bentley-Knight electric line in Allegheny, Pennsylvania.

Whitney liked what he saw, but Longstreet was concerned about the capacity of the electrics to start up a large group of cars that had become bunched during a short section of track, something that happens frequently on a big city system. Sprague decided to show Whitney electric traction's overload capacity. One night, after regular operations had ended, 22 cars were lined up one after another alongside the Church Hill car shed, and the engineer at the power plant was instructed to load the feeder safety-catch fuses, raise the voltage from the usual 450 volts to 500 volts, and to hold on "no matter what."

Whitney and his party were taken from their hotel to Church Hill to witness the test. With the wave of a lantern the 22 cars started up, one after another. The line voltage dropped to scarcely 200 volts, and the car lights dimmed until they were hardly visible, but the cars kept moving. Gradually the voltage began to rise again, the lights brighten, and soon all 22 cars had disappeared from sight. Whitney was convinced and started to convert to electric power rather than cable..[24]

Sprague had lost some $75,000 on the Richmond contract, but he had made many times more based upon the reputation of his work. Whitney soon went before the Boston Board of Aldermen to obtain permission for electric operation, and Sprague obtained an initial contract for 13 miles of track and 20 cars on the West End's Brookline division. Other street railway companies and their manufacturers were soon following suit. The rival Thomson-Houston Electric Company, for example, out-bid Sprague for the West End's 14-mile Cambridge division. But Sprague was by far the leader in early street railway electrification. In General Electric's annual report of 1889, for example, the company reported that "of the 200 electric railways now in operation or in

Redrawn for the August 1905 issue of *The Century Magazine*, Frank Sprague and his men demonstrated how the new electric cars could start even a large group of cars that had become bunched.

the course of construction in the United States the Sprague Company has equipped at least half [actually 113], and over 90 percent of the entire number are operated on the general plan originally laid down by the company as the pioneer in this field of electric work and based upon its patents."[25]

Sprague's triumphant electrification of the Richmond Union Passenger Railway in 1888 marked the transition of electric traction from a period of trial and experimentation to one of successful commercial application. Frank Sprague was not the inventor of the electric railway, and indeed in his extensive writings over the years on the development of the electric railway he always made ample mention of the work of others in the field that preceded or paralleled his own. But if the successful development of electric traction did indeed represent the work of many, it was equally true that the disciplined and scientific approach that Frank Sprague the engineer brought to the task, as well as his own inventive contributions to the technology of electric traction, are what produced the first electric railway that really worked well enough to represent a growing transportation concern.

The significance of the Richmond installation was that it provided the technical features that would permit successful operation of electric street railways on a large scale. These included: the use of a single overhead conductor with the current collection by an under-running trolley; motors on each car, mounted below the car and suspended by an arrangement that made it possible for motor and axle movement without misaligning the gearing; and motor brushes that were fixed in position and motor control that was achieved

One of Richmond's first trolleys was photographed along a downtown street in about 1889. *O. R. Cummings Collection.*

by a series-parallel controller. One controller was mounted on each end of the
car so that it could be operated in either direction. Lightning arrestors were
installed. The current supply system employed 450 volts, a parallel circuit, and
a main conductor which fed the trolley wire at 1,000-foot intervals. All of these
would become standard features that were part of the coming development of
electric traction.[26]

Despite his early success in electric traction development, the Sprague
firm was a small and seriously under-capitalized enterprise in a period when
the growing electrical manufacturers were rapidly expanding and consolidating their business. The Van Depoele firm was first offered to Frank Sprague,
who turned it down, then a few years later it was sold to the Thomson-Houston
Electric Company in 1888. Van Depoele went to work with the Thomson-
Houston's engineering staff, where he continued until his untimely death in
1892. In 1889 the capitalization of the rapidly expanding Sprague firm was

Richmond Union Passenger Railway car No. 19 made a turn near Virginia's state capitol in about 1889. *O. R. Cummings Collection.*

increased by $200,000 in common stock and $400,000 in preferred stock, with all of the latter sold to the Edison Electric Light Co. Shortly afterward the Edison General Electric Company was formed by consolidating several Edison companies. By this time Sprague's traction motor work represented a good two-thirds of all the motor work done by Edison General Electric and, at least in part to be sure that they retained the business, the Sprague firm was merged with Edison General Electric. Sprague did not want to give up his firm, but Edison General Electric's superior financial strength and its position as manufacturing contractor for Sprague won out.

Only three years later the Edison General Electric and Thomson-Houston firms were consolidated to form the giant General Electric Company (GE). This, together with the Westinghouse Electric & Manufacturing Company formed in 1886, completed the formation of the two principal electrical manufacturers that would compete for the American market throughout much of the twentieth century.

Many of Sprague's key engineers went on to work in the newly consolidated Edison General Electric and the later GE firm. Oscar Crosby went on to a long and diverse career, becoming a general superintendent of GE and publishing one of the early textbooks on electric railway practice, and then working in exploration in Asia and as an assistant secretary of the treasurer and other World War I posts. Greene had risen to general sales manager of GE at the time of his sudden death in January 1900, when he and his wife were both drowned in a skating accident on the Mohawk River. In 1891 Ayers Lundy went on to found the still extant engineering and consulting company of Sargent & Lundy which now builds electric power plants on a worldwide basis. Horace Parshall joined the Edison General Electric organization briefly, and then went on to establish a well-known consulting engineering business in London, England.

William Le Roy Emmet, an 1881 graduate of the naval academy who had joined Sprague in 1887, working on such Sprague projects as Harrisburg, Cleveland, and Wichita, Kansas, went on to join Edison General Electric in 1891 and spent a long and distinguished career in developing large-scale GE steam turbines. Like most who had worked with Sprague, Emmet remembered with great praise his chance to work with him. "There are few men in the electrical industry who have been more uniformly right than Frank Sprague and I have always had a great admiration for his intelligence, originality and courage," wrote Emmet in his 1931 autobiography.[27] Frank Sprague himself had agreed to serve as a consulting engineer for Edison General Electric, but it did not work out well for him.

After having developed the electric railway into a highly successful form, Sprague was understandably much opposed to the announcement that Edison would largely supplant the Sprague overhead system, replacing it with running rails at a very low potential, and would substitute a new traction motor of his own design for the one that Sprague had developed so successfully. Sprague was also very upset by the actions of Edison General Electric which would remove his name from any of the equipment he had developed and replace it with the Edison name. In a surviving glossy colored 1891 brochure of Edison General Electric traction equipment, someone—very possibly Sprague himself—has gone through the pages, carefully inking out the Edison name and substituting Sprague wherever appropriate.[28] The magnitude of Sprague's unhappiness is suggested from a few pages of the very lengthy letter of December 2, 1890, by which he resigned his consulting post:

Not only Mr. Edison's subordinates and those who bask in the sunshine of all their smiles, but Mr. Edison himself, forgetful of his dignity and jealous of any man who finds in the whole realm of electric science a corner no matter how small not occupied by himself, loses no opportunity to attack and to attempt to belittle me.

Family quarrels are undignified, and covert attacks are cowardly. I do not care to be mixed up in the former, and I will not remain quiet under the latter. I neither fear Mr. Edison's criticism nor seek his approval. He is indebted to me quite as much as I can possibly be to him. If any attack on me is advisable or necessary let it be made in a manly and open fashion. I am able to reply to it as befits my reputation and dignity.[29]

It was the first open disagreement between Sprague and Edison in the long history of their complex relationship, but it would not be the last.

Now well out of the street railway business, Frank Sprague was already off on another.

5

SPRAGUE AND THE ELECTRIC ELEVATOR

As American cities grew rapidly during the nineteenth century, so did the need for a better form of urban horizontal transportation. That need was finally met by the development of the electric street railway thanks to the work of Frank Sprague and others during the 1880s. And much as the need for improved horizontal transportation was met through the application of electricity, so was it through the use of electric power that the development of increasingly tall buildings was made possible, to satisfy the needs of the growing density of urban cities. Once again, Frank Sprague played an important role, this time in the application of electricity to vertical transportation.

Hand-powered elevators or hoists had been around for centuries, but it was not until the mid-1850s that the modern elevator began to make its way into common use. In 1852 a Yonkers, New York, manufacturer, Elisha Otis, developed a simple and innovative apparatus that provided a reliable and automatic safety device for vertical hoists. The most common accidents with these lifts were the breaking or failing of the lifting rope, or an overloaded platform. Otis fitted notched vertical hardwood guide-rails alongside each elevator. A flat-leaf spring was fastened to the roof of the elevator car and connected to the

In 1891 the electric elevators designed by Charles Pratt and Frank Sprague were selected for the splendid new 14-story Postal Telegraph-Cable Co. in New York, the first major use of electric elevators. *The new building is shown in a drawing from City Hall Park in the April 21, 1894, issue of* The Electrical World.

hoist rope so that it was under tension and arched, and the end springs were drawn in. If the rope broke or otherwise failed, the spring would immediately flatten and shoes on the end of the spring would be driven into the notched rails, bringing the car to an immediate stop.[1] Otis was persuaded to bring his device to the 1854 New York World's Fair, where showman P. T. Barnum had Otis climb onto his elevator, raise it high above the ground and then sever its lifting rope. The automatic safety kicked in, and Otis stood right where he was in perfect safety to the amazement of spectators.[2]

Otis's Union Elevator Works was soon doing a good business in the manufacture of elevators with safety hoists, and in 1857 what is often regarded as the first commercial passenger elevator was installed by Otis's company in the new five-story building of E. V. Haughwout & Company, tableware merchants at Broadway and Broome Street in New York.[3] The Haughwout machine was not much more than a freight elevator, and a much more elaborate passenger elevator was installed in 1859 by Otis Tufts, a Boston machinist and inventor, in New York's splendid new seven-story Fifth Avenue Hotel along Broadway and Fifth Avenue, between 23rd and 24th streets.[4] In any event, the passenger elevator for commercial buildings, hotels, and the like soon grew into a substantial business. Until now, buildings had been limited to no more than about five or six stories in height, but with the advent of elevators the height of buildings began to increase, and well before the end of the century "skyscraper" or "elevator buildings" became ever-more common in New York and other major cities.

This changing landscape was noted by a writer of *New York by Sunlight and Gaslight* in 1886:

> This growth of New York thus illustrated in height is attributed by the architects to the high price at which each foot of real estate is held all over the island, and notably in the lower sections of the city; but it has also been greatly facilitated by the use of elevators, which enable some of the most prominent firms to occupy offices on the fourth and fifth floors, and even higher floors, where only a few years ago they would not entertain the idea of asking their customers to call upon them above the second story. This "mania" for high buildings, which the architects as yet regard only in its infancy, is, however, not original with New York; the new part of the city of Edinburgh, in Scotland, being full of buildings ten and eleven stories high.[5]

The earliest elevators typically used a power-driven drum that moved one or more hoist ropes up and down in the shaft, with one end attached to the elevator car, and the other to a counterbalance. Early elevators used a hemp rope, but this was soon replaced with the much stronger and more reliable wire rope cable developed by John Augustus Roebling around 1840. The hoist ropes were attached to a spiral-grooved winding-drum, and passed over it to be wound onto or off of the drum as the car traveled up or down, while the hoist rope attached to the counterweight moved in the opposite direction. The winding-drum was originally powered by a steam engine, but by the mid-1860s hydraulic machines were being used as well.[6] Many of these used a long piston or plunger, usually in a long cylinder under the elevator, while later hydraulic designs used horizontal or vertical pistons and a combination of sheaves and ropes to multiply the car motion.

The winding-drum and the hydraulic elevator were limited in both the maximum feasible height and speed that they could use. By the beginning of the twentieth century, a newer elevator design, usually called the gearless-traction type, used a directly attached power motor to drive a grooved sheave over which the hoisting ropes transmitted the necessary lifting power through the friction between the sheave and the hoisting ropes. This new arrangement permitted both much higher hoisting height and operating speeds.

Using cast or wrought iron beams and columns, and masonry load-bearing walls, New York buildings by 1875 had reached heights of as much as 230 to 260 feet. Within another decade the design of buildings using an all-steel framework made possible the construction of even taller buildings. New York's steel frame Tower Building, designed by architect Bradford Gilbert and completed in 1889, reached a height of 11 stories, and still taller buildings soon followed.[7]

Well into the 1890s, Otis and other elevator manufacturers continued to use steam or hydraulic equipment. But the rapidly advancing technology of electricity would soon come to the elevators as well. In 1881 Siemens had exhibited an electric elevator at the Paris Exhibition. The car was partially supported by counterweights, with a motor mounted under the elevator car, but the car design was not put to commercial use. Several years later electric motors began to be used for powering drum-type freight elevators. Electrical inventor Leo Daft installed one in the Garner Cotton Mills in Newburgh, New York, in 1883, and a year later Frank Sprague used one of his electric motors to power a freight hoist at the Pemberton Mills in Lawrence, Massachusetts. This was the first example of an elevator operating at uniform speeds despite variable loads and with regenerative return of energy to the line when an unbalanced load was moving in the same direction of movement as the elevator.[8] In 1886 a Boston company installed a 15-horsepower, 220-volt Sprague motor in a building at Purchase and Pearl streets in Boston to power a freight elevator.[9]

There were several electrical inventors who developed ideas for electrical passenger elevators. Engineer William Baxter, Jr., had begun work on a direct-connected elevator as early as 1884, and a direct-current, worm gear machine was put in service in Baltimore in 1887. While still building only steam and hydraulic machines, the Otis Elevator Co. asked its chief engineer, Thomas E. Brown, to look into electric operation of its elevators.

The first electric elevator work for Otis was developed by Rudolf Eickemeyer, a Bavarian-born engineer who also lived in Yonkers. Eickemeyer had already developed, together with Steven D. Field, an experimental electric locomotive design that was tried out on the Manhattan Railway in 1887, and a novel design for a street railway car. Eickemeyer turned his attention to electric elevators and in 1888 Otis bought shares in the business. The first attempt to build an electric elevator combined an Eickemeyer streetcar motor with an Otis worm gear elevator machines. In 1889 two of these were installed by Otis in the Demarest Building in New York, where they proved to be a solid success. In 1892 Otis bought out Eickemeyer and renamed it the Otis Electric Company, adding electric operation to the leading elevator company.[10] There were soon other competitors for electric elevators.

With his electric railway business sold out to the Edison General Electric Co. in 1890, Frank Sprague was ready for new electrical challenges, and the development of the electric elevator looked like an opportunity with potential. Already at work on a promising design for an electric machine was Charles R. Pratt, a mechanical engineer who had graduated from the Massachusetts Institute of Technology and had secured several patents on elevators. In 1888 Pratt had begun the design for an elevator that would combine the mechanical operation of the hydraulic elevator engine with the more efficient operation of the electric motor. Pratt designed a high-speed screw electric elevator that used a Thomson-Houston electric motor with the horizontal frame from a Whittier hydraulic machine. Frank Sprague joined with Pratt in 1890, and they organized the Sprague-Pratt Electric Elevator Company, soon changing the name to the Sprague Electric Elevator Company on June 13, 1892. Edward H. Johnson was again, as he had been for the formation of Sprague's electric motor business in 1884, one of the officers of the new company.

Further work on the Pratt design was made to ready it for production, and a prototype was installed in a building at 135 West 23rd Street in New York in 1891. While the design was primarily Pratt's—it was patented by him in 1891—Sprague also contributed significant work to the development of electrical controls and safeties for the machine, and his financial support and work in marketing and promotion were key measures in getting the new company started.[11]

Sprague and Pratt landed their first elevator contract for the small Grand Hotel at 33rd Street and Broadway, and then, in 1891, a major contract for the electric elevators of the splendid new Postal Telegraph-Cable Company at the corner of Broadway and Murray Street in New York. This was a daunting task for the new elevator company. Sprague and Pratt had to duplicate the first-class service in speed, control, safety, and capacity provided by a modern hydraulic elevator, to improve upon some of its characteristics, and to occupy less space. At the same time, the electric elevators would have operating costs not greater than one half those of a comparable hydraulic plant. Sprague got the contract, he later said, only by assuring John W. Mackay and George W. Harding, the building owner and architect, that he would replace the electrical elevator system with a hydraulic one if his work proved unsatisfactory. But the confident Sprague predicted that "he would write the epitaph of the hydraulic elevator as that of the horse car had been written at Richmond."[12]

The 14-story building was of particular interest, for it was to be the first major building equipped with electric elevators. Four elevators were local or "way" elevators, serving floors 1 through 11 and a restaurant on 14, while two express elevators served only floors 1 and 11 through 14. The operation of the elevators was an arrangement similar to those of horizontal and vertical hydraulic machines. The four-pole direct current motors were linked through wormed gearing to sheaves and ropes in a horizontal arrangement, in which the hoisting cables wound around the multiplying sheaves and then ran to the top of the shaft, and then down to the shaft, with the counterweight attached to the car by a separate set of cables. Multiplying sheaves were introduced on the shaft, with four to six cables run over a set of sheaves at the top of the shaft, and down to a combined sheave-and-counterweight assembly and then back to the top of the shaft. The upper sheaves doubled the multiplying action of the horizontal engine, requiring fewer sheaves and reducing the cable lengths. The arrangement of the elevators was such that the motor always took current from the line in hoisting, but was always cut off from the line and operated as a driven dynamo by the falling weight of the car. The machine was capable of speeds as high as 600 feet per minute.

Frank Sprague claimed that this arrangement was the best one yet designed for long rise buildings, providing both much less and more uniform wear on the hoisting cables. But with all of the new ideas and equipment used on the Sprague-Pratt elevators, getting it all into place and working well proved

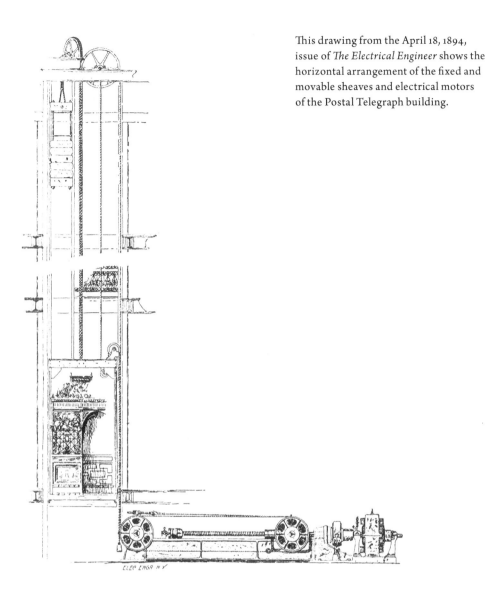

This drawing from the April 18, 1894, issue of *The Electrical Engineer* shows the horizontal arrangement of the fixed and movable sheaves and electrical motors of the Postal Telegraph building.

to be a long and difficult task. During the installation of the elevator plant, Frank Sprague and several of his men came uncomfortably close to a serious accident or death with the just-completed first run of an elevator machine.

On this occasion, having late in the evening made a successful individual run with the first machine, I in an elated frame of mind directed all the crew save one man to join me. Pilot motor control had been adopted, but on this ascent the car could not be stopped, and as neither the upper limit nor the car safeties had been installed there flashed an unpleasant vision of heading into the overhead sheaves at nearly 400 feet a minute, the snapping of the cables and a 14-story free drop, with a tangled mass of humanity and metal the object of a coroner's inspection. This was the picture, and the only one

A photograph of the Postal Telegraph building from the April 21, 1894, issue of *The Electrical World* provides a close-up view of the hoisting nut and sheaves.

in mind as without a word to my men and standing rigidly I vainly operated the controller. When two feet above the last floor and with but two more to the limit, the car suddenly stopped, the sole man left in the engine room having had the wit to pull the switch, which cut off the current and operated the electro-mechanical brake. We walked down, but the car followed leisurely under control.[13]

Making repairs and improvements proved difficult as well. "This Postal Telegraph experience was a somewhat wakeful one also," recalled Sprague, "for every machine had to be rebuilt while in operation, and for weeks it was doubtful whether the outcome would be a success."[14]

The Postal Telegraph experience was also a proving ground for new devices developed by Sprague and Pratt. Among the new advances developed by the two men over the next several years, many having been tried first on the Postal Telegraph system, were pilot motor and push button control, electro mechanic brakes, dead man's switches, and automatic self-leveling. By far the most important and far reaching of these new advances was the concept of pilot control of multiple electrical machines. In 1895 Sprague had made arrangements for the interconnection of controlling circuits in the Postal Telegraph building's basement, with provision to throw any or all machines onto a common master switch and run a number of elevators simultaneously from the single switch. While this "multiple unit" concept had some useful characteristics for the electric elevator, Frank Sprague quickly realized that it would be a much more important concept for multiple unit operation of electrified railroad cars, and it would soon take Sprague back into the electric railway business.

The innovative design of the Postal Telegraph elevators attracted much interest from other engineers, and Frank Sprague made an extended presentation on the system to members of the American Institute of Electrical Engineers at a New York meeting of the institute on January 22, 1896. Clearly, he concluded, the Sprague-Pratt multiple sheave electric screw elevator was superior to the old hydraulic machines.

"The net result has been," said Sprague, "that this machine now stands the superior to the hydraulic elevator in that it has its speed and capacity with, if anything, greater safety, and certain advantages in its automatics."

"On high lifts it occupies less space; it is more flexible in its application, is more economical to operate, and it is more easily cared for."[15]

Not everyone at the meeting agreed. Engineer George Hill, for example, stood firmly on the side of the high-speed hydraulic elevator.

"I have ridden on a great many high-speed hydraulic passenger elevators," he said, "and, although I am a very strong advocate of electric elevators in their place, I affirm that so far as my experience goes, there is absolutely no machine to-day which can compare with the hydraulic high-speed passenger elevator."[16]

The discussion of Sprague's paper was so lively that it was continued for a second meeting a month later, on February 26, at which a number of readers discussed such questions as the impact of high starting motor surges on electric systems, comparative operating costs of electric vs. hydraulic machines, and Sprague's response to their questions.[17]

Engineers were intrigued by the new elevator designs of Sprague and Pratt. Chicago consulting engineer Louis K. Comstock remembered with much amusement many years later a visit to one of the new designs. He "regarded Sprague with awe" when he came to New York in 1893 to look over the Sprague elevator at the Grand Hotel for an architectural firm:

Since I had come all the way from Chicago to see this piece of electrical mechanism, he would show it to me—guts and all. Between 12 midnight and early morn it was skinned, dissected, dismembered and held up to the light;

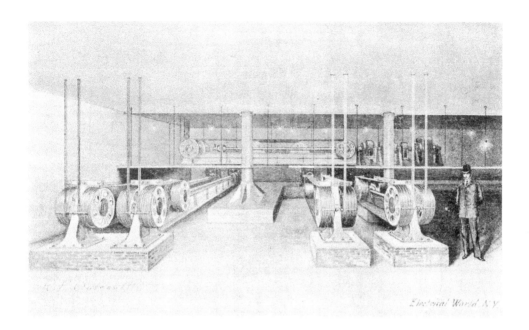

Artist R. F. Outcault's general view of the Postal Telegraph building's electric elevator plant and a close-up view showing the arrangement of an electric motor and brake. *Drawings by R. F. Outcault for the April 21, 1894, issue of* The Electrical World.

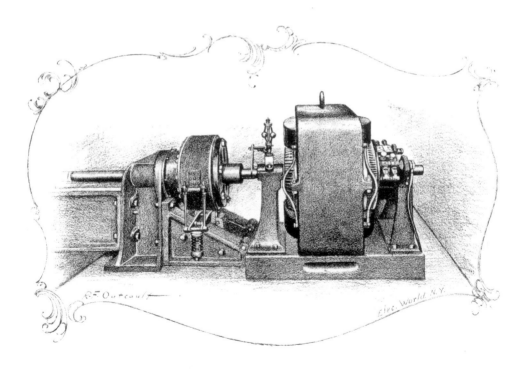

By 1894 Frank Sprague had completed the construction of a major new plant, with almost 300,000 square feet of space in Watsessing, New Jersey. This photograph shows the exterior view of the new building. *Frank Sprague Papers, Manuscripts and Archives Division, The New York Public Library, Astor, Lenox and Tilden Foundations.*

put under the microscope, subjected to the acid test, and put together again and survived. This major operation was performed on the patient while I took the anesthetic administered by Dr. Sprague and his anesthetist, Dr. Pratt. The anesthesia was a constant flow of smooth, easy, oily jargon which completely hypnotized the visiting nurse.

I marveled that they didn't have a score of pieces left over when the synthesis was completed. How Dr. Pratt accounted for all the parts and made them look as though they had been put together again is to this day a mystery to me.

Dr. Sprague assured me that the traveling nut would not overrun its travel because it was the physical concept of the square root of minus one; whereupon Dr. Pratt, leaning heavily on Charlie Benton, gravely said "these case hardened steel balls perform the function of the differential co-efficient of the speed of the screw times the radius of gyration of the unknown quantity to the nth power." And Benton said, "l'enfant dit vrai."[18]

Sprague had initially begun his elevator work on a rented building in Manhattan. In 1893 the company began to build its first plant in Watsessing, New Jersey, between Montclair and Orange, and began operation a year later. This was soon outstripped by the growth of Sprague Elevator and several years later the company added a new plant for the growing work. The six-story, almost

This interior view of the new Watsessing plant shows the main assembly building with its overhead crane.   *Frank Sprague Papers, Manuscripts and Archives Division, The New York Public Library, Astor, Lenox and Tilden Foundations.*

300,000-square-foot plant was built of steel and concrete, with brick exteriors and extensive windows, making the proud builder say it "would be hard to imagine a more perfect structure." The plant was capable of employing more than 700 workers, with special arrangements made to accommodate women, who were expected to wear long knickerbockers, with snug-fitting caps and close-fitting sleeves to protect them against machinery.

Well before the end of the decade, the Sprague Electric Elevator Co. had grown into a comprehensive company, offering a wide range of electric elevators and hoists, equipped with both screw and geared winding drum machines. Elevator cars were available in a wide range of shapes and capacities. Cars could be controlled either by an operator or automatically through push buttons operated by the passenger, and Sprague developed a wide range of safety devices for the cars. Both passenger and freight elevators and an automatically controlled dumb waiter service were also available.[19]

Competing with such giants of the elevator business as Otis and A. B. See, Sprague managed to land several prestigious contracts. In 1896 Sprague got the commission to build the elevators for New York City's Park Row Building, designed by architect R. H. Robertson, which was then the world's tallest building. Sprague and Pratt had developed a new vertical screw system for the ten passenger elevators for the 31-story building. The motor and screw shaft for each elevator was mounted vertically at the bottom of each shaft, while the sheaves were installed in the elevator shaft, with the fixed sheave located near the 16th floor of the 297-foot shaft. Instead of being connected to the elevator car, the hoisting cables were instead linked with a traveling counterweight/sheave assembly, which in turn was connected to the car.[20]

In 1897 Sprague and Pratt won the contract for a nearly $500,000 project, the largest elevator contract yet awarded anywhere. This would provide the elevators, or "lifts," for the Central London Railway, one of London's earliest undergrounds. The 5.7-mile subway would serve a total of 14 stations between Shepherd's Bush and Bank. The entire electrical equipment for the underground was built by American suppliers. Sprague, of course, built the elevators, while all of the line's traction equipment was supplied through British Thompson-Houston Company, representing the General Electric Company of Schenectady. The underground locomotives were B-B type, steeple cab locomotives weighing about 45 tons, while direct current power was supplied through a center third rail located between the running rails.

The deep tunnel subway required an installation of 48 elevators, each with a motor capacity of anywhere from 100 to 150 horsepower and capable of handling an unbalanced load of up to 17,000 pounds. These were located in 24 shafts, with rises of anywhere from 41 feet to 91½ feet. The electric motors were powered by double worm gear machines that powered winding drums rather than the multiple sheave electric screw elevators used by Sprague and Pratt on the high-rise buildings in New York. The engineering work for the machines was done largely by Charles Pratt and Job Rockfield Furman, an experienced elevator consultant who had previously worked with Otis.[21]

Sprague gained the Central London work only after a long and uncertain period. Bids had been solicited from several elevator companies, with each company sending a representative for the bidding. All had proposed to develop the elevators with a hydraulic system except Sprague, who of course proposed an electric system. Sprague himself would represent Sprague Electric Elevator.

At the same time he needed to get work started on a very demanding project for electrification of a Chicago elevated railway project. Sprague got some preliminary design work done and agreed to Chicago's demanding time schedule before heading to New York. Then, badly injured in an accident with an elevator installation for the Waldorf Hotel in New York, he was still on crutches when he left for London.

The engineers for the Central London—Sir John Fowler, Sir Benjamin Baker, and Mr. Basil Mott—completed their evaluation of the competing proposals, but it would be another month before they would reach a decision. While Sprague had offered a tender in both price and time that was substantially under those for their British competitors, Central London's conservative engineers were uncertain about contracting for the new electric machines rather than the familiar hydraulic machines. Although anxious to return to New York for the Chicago project, Sprague had to repeatedly delay his departure from London until the contract award was settled. Finally, Sir Benjamin Baker,[22] as the principal engineer for the subway, agreed to go ahead with the Sprague-Pratt electric elevators, but only on the condition that a test installation by the subway's engineers be done to satisfy them that the machines could meet their design criteria in all respects. Only then would the bid be awarded. These design criteria included demanding standards for the speed and frequency of the elevators. In addition, for electric elevators, with an average total of 20,000 single trips per day with a 67-foot mean rise and the then-current one penny per kilowatt hour rate, the elevators had to work for three years at a cost not exceeding one pound per 1,000 trips. This would represent a cost of less than half that for a hydraulic elevator under similar conditions.

Confident that they could meet the subway's criteria for the contract, Sprague sailed for New York in mid-June. Within about four months, a working test installation was completed in an 18-foot-diameter shaft at Notting-hill-gate for an exceptionally rigorous and unusual series of tests conducted by the Central London engineers. The results were fully satisfactory in every respect. Sprague's men were so confident of their ability to meet the criteria that they had already started work on the remaining 47 shafts even as installation for the initial working test was still being completed. Heavy overhead work for the project was supplied by the Carnegie Company, while the balance of the operating machinery was supplied by Sprague's Watsessing plant. Only the required steel cables were supplied from an English firm.[23] Completion of the entire project was required in 16 months from the date of contract award.

But even as he successfully competed for the largest single elevator project ever carried out, Frank Sprague had decided to leave the elevator business.

This may have been partly about his future prospects in that business, but perhaps equally convincing were the prospects for his major re-entry into the electric railroad business. Between 1892 and 1898 Sprague and Pratt had supplied a total of 584 elevators. While this was a respectable performance, it fell far behind such giants as Otis. Just between 1884 and 1896, for example, 360 electric elevators powered by the New York Edison power system were installed in New York City. Of this number Otis had built some 200 elevators, while Sprague had installed only 13. Perhaps even more important was Otis Elevator's aggressive move to dominate the elevator business. With the incorporation of 15 other companies under the Otis umbrella in 1898, the company conducted about 90 percent of the U.S. elevator business. Sprague sold Sprague Electric Elevator Co. to Otis even before the Central London project was complete.[24] But in any case, Sprague had played an important role in the shift from steam and hydraulic power to electricity for elevators.

THE DUAL ELEVATOR SYSTEM

As America's principal cities continued to grow in the twentieth century, the need for more space in urban centers and the rising cost of land inexorably pushed the height of skyscrapers upward. At the end of the nineteenth century, the tallest building in the world was the 31-story Park Row Building in New York City (for which Frank Sprague had supplied the elevators). By 1907 New York's Singer Building had reached 47 stories, and in 1913 the Woolworth Tower had climbed to 55 stories. The Chrysler Building of 1930 would reach 75 stories, and only a year later the completion of the Empire State Building would reach the world's tallest building height of 85 stories. The Great Depression and World War II brought a temporary halt to the construction of still taller buildings.[25]

As office buildings grew taller, the amount of building space that had to be reserved for elevators steadily grew as well, which meant the loss of rentable area. Each elevator operating to the top of a 100-story building, for example, required about 6,000 square feet of floor area that could otherwise be rented. There were several approaches to reducing the space required for elevator shafts. As buildings grew taller, the elevators were divided into express and local cars. The need for elevator space could be further reduced through the use of automatic stopping and leveling devices, which operated more rapidly, and the use of faster elevator motors. By 1930 elevator car speeds of up to 1,200 feet per minute were being planned. Several elevator manufacturers came up with novel two-story elevators, which placed a second car above the first. Passengers would board the cars at two levels, with one car serving even-numbered floors and the other odd-numbered floors.[26]

Still another option for increasing the capacity of elevators was Frank Sprague's concept of a dual elevator system, which would permit the operation of two—or more—elevators in the same shaft. Although by that time Sprague had been out of the elevator business for more than 25 years, a prominent New York real estate operator remembered his pioneering work with electric elevators. Realizing the substantial increase in rental income that would be possible with a dual elevator system, he asked Sprague about its feasibility. By the end of 1926 Sprague had completed an outline of how the system would operate and had completed a working model. The heart of the system was Sprague's use of electrical controls that prevented the two elevators from entering the same space.[27]

Sprague submitted his patent application for the dual elevator system late in 1926, with the patent approved in 1930. Sprague sold the patent to the Westinghouse Electric & Manufacturing Company. Two Westinghouse engineers, Henry D. James and Walter S. Rugg, incorporated some additional features and equipment, and the first trial Westinghouse-Sprague Dual Control elevator was installed in a new 11-story Westinghouse main office building in East Pittsburgh, Pennsylvania, early in 1931.

The typical arrangement developed by Sprague placed an upper express elevator and a lower local elevator in the same shaft. The 11-story Westinghouse building was divided into a lower local car serving 9 floors and a top express car serving 3 floors. The controls at the top of the upper elevator and those at the bottom of the lower elevator would function much like those of an ordinary elevator. The controls between the two elevators, however, would be in a "floating" control zone, shifting automatically between floors based upon the movement of the two elevators. With the floating control, the upper elevator would be controlled at the bottom, and the lower elevator controlled at the top, with the limits of travel determined by the distance between the two elevators. When the two elevators were approaching each other, a slow-down zone would be established within a prescribed limit, such as three floors, and a stop zone established when they were one floor apart. Green, yellow, and red lights would inform the operator of the space between cars, with each car automatically slowed down and then stopped by the same control which was used for normal operation. Each car in the system would always maintain contact with the system's block control circuits with moving brushes through metal strips mounted in the shaft walls. Three systems—one electrical and two mechanical—would ensure safety.[28]

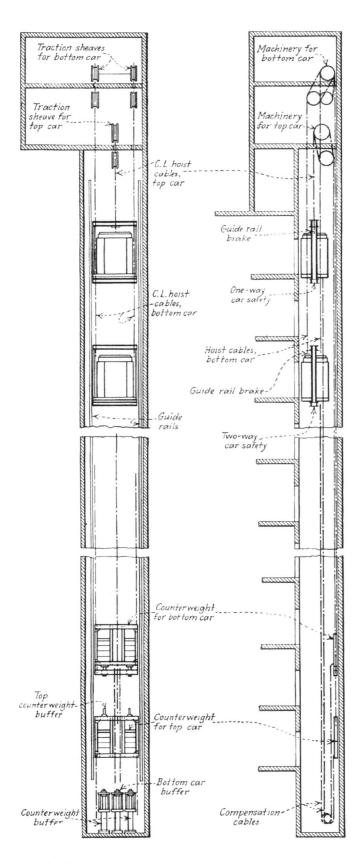

Traction sheaves
for bottom car

Traction
sheave for
top car

Machinery for
bottom car

Machinery
for top car

C.L. hoist
cables,
top car

Guide rail
brake

C.L. hoist
cables,
bottom car

One-way
car safety

Hoist cables,
bottom car

Guide rail brake

Guide
rails

Two-way
car safety

Counterweight
for bottom car

Top
counterweight-
buffer

Counterweight
for top car

Bottom car
buffer

Counterweight
buffer

Compensation-
cables

The first drawing shows
how the innovative
Frank Sprague design
for multiple cars in the
same shaft would work.
The second simplified
diagram of an electrical
block system shows
how two or more cars in
the same elevator were
controlled by the control
circuits.   *February 14,
1931, Electrical World.*

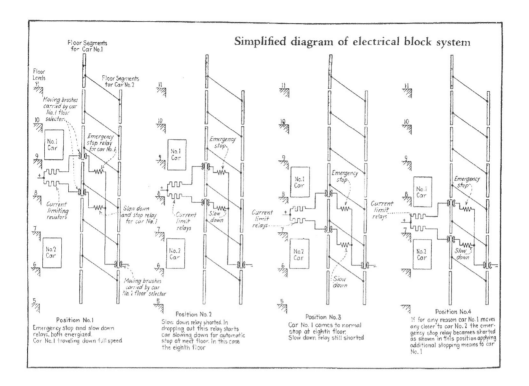

Simplified diagram of electrical block system

The new elevator system attracted much public attention. Technical journals such as *Engineering News, Architectural Record, Power,* and *Electrical World* published details of the Sprague system. Popular magazines such as *Scientific American* and *Popular Science Monthly* and newspapers carried extensive articles. "2 Elevators on One Shaft; Believe It or Not," said one newspaper.

Other manufacturers were also interested in the new technology. Otis Elevator had considered the dual elevator system soon after Sprague developed his design, and two Otis patents for the system were awarded in 1931 and 1933. The 50-story Irving Trust Company Building, completed in 1931, was said to include a provision for the later addition of dual elevators, although this was never done.

In any case, Depression-era 1931 was not a good time to introduce the design of a new elevator system for high-rise buildings, and the Sprague dual elevator design was never repeated. But as "mega high-rise" buildings became increasingly common in the years following World War II, elevator manufacturers were again looking at innovative vertical transportation systems to efficiently move their passengers on extremely high-rise buildings. Always, the objective was to find ways to move people up and down the building more efficiently so that enough rentable office space would be available to make a building economically feasible.[29]

Some new buildings used the idea of two-story or "double-decker" elevators. New York's World Trade Center, completed in 1973, whose two towers were then the world's tallest buildings, introduced the idea of sky lobbies, which were like a transfer station, taking passengers up and down on an express and then transferring them to or from a local at the sky lobby. Recently elevator manufacturers have developed computer-controlled "destination dispatch" or "smart elevator" systems, which assign passengers to an elevator according to their floor destination, reducing each car to as few stops as possible.

Even now, ever taller buildings are being developed. The Taipei 101 building in Taiwan, which became the world's tallest in 2004 with a height of 1,667 feet and 101 floors, operates its double-deck, pressurized elevators at more than 3,300 feet per minute, qualifying them as the world's fastest. By late 2009 the even taller Burj Dubai in the United Arab Emirates will be completed, with a building at least 2,684 feet high, and still faster elevator systems running at maximum speed of more than 3,500 feet per minute. Still taller buildings being planned in the Middle East and Japan could raise building heights to as much as 3,900 feet. And so, elevator manufacturers will be planning how to make them work. Perhaps one day elevator engineers will again be taking a look at Frank Sprague's imaginative idea of two elevators in the same shaft.[30]

The first major electrification of an elevated railway was the Columbian Intramural Railway, which operated on the grounds of the World's Columbian Exposition at Chicago in 1893. A system for controlling separate electric cars in the same train was not yet available, and the Columbian trains employed a single combined passenger and locomotive car, pulling three unpowered trailers.   *Duke-Middleton Collection.*

6

FRANK SPRAGUE
AND THE MULTIPLE
UNIT TRAIN

In 1890 when Frank Sprague turned his attention to the development of high-speed electric elevators, he had by no means given up his interest in electric railroads. As early as his 1882–1883 visit to London, he had developed ideas for the electrification of the city's steam-powered subway. In 1885 he had developed and presented a plan for the electrification of New York's steam-powered elevated lines, and by 1886 had developed and tested electric equipment for the Els. With the elevated companies still not interested in electrification, he had turned his attention to street railway electrification, which led to his great success at Richmond and a boom in electric railways in the latter 1880s. By 1890 he was pursuing opportunities for electrification of elevators.

But Sprague did not set aside his interests in rapid transit for long. By 1891, New York's Board of Rapid Transit Commissioners was struggling with the problem of extending the already existing elevated railway system or beginning the development of an entirely new subway system. Frank Sprague, in a long interview with the *Commercial Advertiser* on February 16, 1891, spelled out how he believed the city could best solve the increasing urgency of an expanded transit system. New York, he said, should build a four-track independent way and express tunnel service, using electricity as a motive power, which he assured the reader would "be capable of satisfying in the highest degree the most exacting demands of the service."[1]

Just a month later, on March 15, Sprague addressed a letter to William Steinway, chairman of the Rapid Transit Commission, advocating the use of an underground system of tubular construction, and adoption of electricity as a motive power. "I am ready," he said, "if a rapid transit system be adopted requiring the use of the electric motor, to undertake the entire contract for the necessary steam and electrical equipment for not less than fifty way and express trains operated as I have outlined, under satisfactory guarantees of efficiency and cost of operation as compared with steam practice."

"I repeat," he emphasized in the close of his letter to Steinway, "there need to be [sic] no hesitation on the part of your Board on the question of electric traction because of any apprehension that the electric motor development will be found wanting when demanded."[2]

In a May 21 presentation to a general meeting of the American Institute of Electrical Engineers, Sprague reviewed the options for elevated versus subway construction, and steam versus electric power available to the Commission. At least the use of some underground construction was clearly in New York's best interest, he advised, and—to no one's surprise—he advocated the use of electric power. Competing rapid transit systems, he suggested, could be selected on the basis of the merits of the alternative systems, considering the best and fastest construction, the best offer for the franchise, and the highest financial guarantees for the satisfactory performance of the system.[3]

Whether by means of a new subway or by electrification and expansion of the elevated railways, New York's Rapid Transit Commission was finding the building of rapid transit to be an exceedingly difficult topic. Confronted with the enormous potential cost for building a new subway system, the Board had struggled for nearly a decade without finding an acceptable solution for financing it, while the city's privately owned elevated railroad companies much preferred to extend the elevated roads. And so far as Jay Gould, who controlled the elevated roads, was concerned, "there is no electric motor in existence, or likely soon to be invented, capable of drawing heavy trains rapidly and economically over the required distances."[4] So all of Frank Sprague's earnest proposals of 1891 fell on deaf ears, and nothing happened.

Two years later, Sprague was back again with a new proposal. The Commission had finally favored a subway powered by electric power, but there were a number of critics, representing what Sprague called "existing corporate interests," who attacked the idea of electric power. Some said, for example, that no electric motor had yet been invented that could propel even one loaded elevated car at 30 miles an hour. Sprague suggested that there should be an actual practical test on a large scale so that electric motors could demonstrate that they could do what was claimed for them. "I am entirely ready to make a demonstration," Sprague proposed, with his usual confidence in his work.

The test section proposed by Sprague was to be a two-mile section, operated by trains of six standard elevated cars. One would operate with a separate electric locomotive, and one with a system of motors, one on each car, all controlled from a pilot locomotive. Operating conditions would include a maximum speed of not less than 40 miles per hour, and the power to stop a train without the use of brake shoes and independently of the main station current. These requirements of both speed and weight were double those already characterizing London's new electrified underground, Sprague pointed out. The cost of making the test runs would be paid by the Commission to whoever won the franchise for the new subway, while if the test failed, all costs would be borne by Sprague. Despite all these proposals, once again, nothing happened.

Sprague had developed a concept for what he called "distributed motive power"—with multiple distributed motors powered from a single control—as far back as the time of his December 10, 1885, presentation to the Society of Arts in Boston. But he had not yet worked out a way to operate the decentralized multiple motors. When his work on the electric elevators for the Postal Telegraph Building a decade later developed a concept for controlling multiple electric motors through a single master control switch, Sprague knew that it would provide far greater benefits to electric railroads. He had completed early sketch plans for the system by 1895, and he was soon ready to try again with the New York Rapid Transit Commission with what is now called "multiple unit control."

Understanding that the Commission was then considering a new idea of some extensions of the Manhattan Railway's elevated lines, as well as motive power modifications, on June 6, 1896, Sprague wrote them once again. "I shall be glad to appear before your Board to make in a definite manner a proposition either in connection with the Manhattan Company, or entirely independent through myself and some associates, for a serious demonstration on the Ninth Avenue, or some other division" and he went on to outline the general characteristics for the demonstration.

Sprague required a site that had enough double track with sidings and switches, and with as much grade to give the demonstration the severest test. Five existing cars would be modified to provide a traction motor on each car, large enough to operate the demonstration train at express speeds, with each car equipped with automatic control so that the cars would be self-braking in the case of an accident. The key part of Sprague's proposal was that each car would be provided with a special control to permit a car to be operated from either end at will, or for the operation at either end of a train composed of up to five cars in any required combination and without regard to their sequence.[5]

Sprague noted that these distributed motor cars would be able to operate trains more frequently than was possible with locomotives, and that trains could be of any length, and operated from either end. With the use of multiple unit control, trains of any length could be properly matched with the required motive power, while the provision of distributed motive power to all the cars of a train would permit much faster acceleration or deceleration. Substantial cost savings could be made in the company's coal account, and a long list of other improvements would be possible. Elimination of locomotives would eliminate the need for head and tail switching, increasing the speed of dispatching trains, and reduce the wear on the elevated superstructure and tracks. Provision of electric power would permit better lighting of trains. All of these would provide a marked increase in the number of passengers carried.

"In short," Sprague concluded, "a very large return on the capital required for a change of motive power. On completion of this demonstration I shall be prepared to make a bid for the electric equipment of any part or the whole of the present or extended system in its entirety."

This proposal, too, aroused no interest from the rapid transit board. Sprague tried one more time in early 1897. In a February letter addressed to a special committee of the Manhattan Elevated Railroad headed by George Gould—who had succeeded his infamous late father, Jay Gould, in 1892—and including financier Russell Sage and R. M. Gallaway, Sprague offered similar arguments.[6] He followed with another letter to George Gould a month later, but once again received no response.[7] "This letter met with a similar fate," wrote Sprague, "and finding my philanthropic efforts unappreciated, I desisted for a time from frontal attacks, and busied myself with the elevator business to which I was financially deeply committed."[8]

While New York's elevated lines remained uninterested in electrification, the lead in electric rapid transit operation would soon shift to Chicago. The fast-growing city began construction of its first elevated railway, or "L,"[9] in 1890, and the system would develop in four separate companies, radiating south, southwest, west, northwest, and north from the central business district, while a fifth company built the downtown Chicago "Loop" that tied all of the lines together. The earliest "L" lines were built with Forney-type steam locomotives, similar to those used in New York. The first one to open was the Chicago & South Side Rapid Transit Railroad Company, which began operating between a Congress Street downtown terminal and 39th Street in May

1892, while service was gradually extending south to 61st Street and then east to Jackson Park by May 12, 1893, just two weeks after Chicago's great World's Columbian Exposition opened in the park. A second Chicago "L," the Lake Street Elevated Railway Company, began operating in November 1893, also with steam power.

While Frank Sprague's energetic promotion of electric power for the New York elevated produced no results, Chicago had decided to proceed with an electric elevated line that would transport fairgoers around the grounds of the Columbian Exposition in 1893. The exposition president, H. N. Higinbotham, had estimated that 30 million people would attend the exposition, with an average daily attendance of 200,000, while transportation planners had estimated that the peak number of visitors arriving at the site might exceed 100,000 an hour. To handle these enormous crowds, the exposition planned to build the Columbian Intramural Railway to serve the 633-acre site and some 200 major buildings in Jackson Park. The elevated railway would extend over a 3½-mile route serving ten stations, as well as making direct connections with stations for both steam railroad trains and the South Side Elevated Railroad.

The fairground was often referred to as the "White City," and made use of the new science of electricity for every possible purpose. When President Cleveland officially opened the fair, he pushed a button and 100,000 light bulbs brilliantly illuminated the white structures. Naturally, the exposition's planners had envisioned that the elevated railway would be electric, even though nothing approaching this scale had ever been built before.

Electrical work for the railway was supplied by General Electric (GE), which had just been formed through the merger of the Thomson-Houston Electric Co. with the Edison General Electric Co. on April 15, 1892. The Intramural Railway was supplied with 72 double-truck cars built by the Jackson & Sharp Company of Wilmington, Delaware. Every fourth car was a "locomotive car" fitted with four 50-horsepower motors, with power supplied from a third rail at one side of the track.[10]

Chicago's Intramural Railway was a huge success for the first major elevated railway electrification. During the 184 days the Exposition was open, from May 12 through October 31, the railway carried a total of 5,083,859 passengers, with a peak ridership of 70,000 to 80,000 passengers per day. During peak periods as many as 15 trains were in operation on the line. Elevated trains were normally operated in four-car trains of one locomotive car and three trailer cars, although on occasion a single locomotive car proved capable of pulling a train with as many as seven fully loaded trailer cars without difficulty.[11]

An aerial view of the Exposition showed a train at the grounds' North Loop, with Lake Michigan in the background. The four-car train was made up of a powered car pulling three trailer cars. *Smithsonian Institution (Neg. No. 2001-3252), National Museum of American History.*

The executives of Chicago's elevated railways were quick to recognize the improved service and the more economic operation that electric power could provide. In April 1892 construction had begun for Chicago's third elevated line, the Metropolitan West Side Elevated Railway, which was planned for operation with steam power. Rolling stock for the Metropolitan was already on order, but based upon the Intramural Railway's extraordinary success with electric power, the Metropolitan's directors changed their contract in May 1894 to allow the substitution of electric power instead of steam. An order with the Baldwin Locomotive Works for 60 steam locomotives was cancelled, and 55 electric locomotive cars were ordered instead from the Barney & Smith Car Company in Dayton, Ohio. Regular service over the Metropolitan's Northwest branch, Garfield Park line, and Humboldt Park branch began from May through June 1895, while the Douglas Park branch followed the next year. Both a higher level of reliability and much-reduced operating costs quickly demonstrated the wisdom of the company's conversion to electric power. By late 1895 the Metropolitan's operating costs per train mile were reported to have declined to 22½ cents a mile, while operating costs for the steam-operated South Side "L" were reported at 48 cents per train mile.

Chicago's Lake Street elevated was the next line converted to electricity. In January 1895 the line decided to convert to electric power, and contracted with a local car building firm to convert 37 of its passenger cars to locomotive cars with General Electric motors and other electrical equipment, and new trucks. All work for the conversion to electric power was completed by April 1896, and operation of steam power had ended by June 13.

By this time only the Chicago & South Side Elevated Railway still operated with steam power. While the line had carried record numbers during the period of the World's Columbian Exposition, it did not do so well following closure of the fair. "L" patronage and revenues after 1983 dropped abruptly, and stock values followed. Whereas most elevated roads were built on structures above public streets, the South Side had decided to build on private right-of-way above alleyways between streets, giving the line its informal name, the Alley "L." The combination or reduced revenues and the capitalized costs for the additional property brought the railway into receivership on October 5, 1895.

A reorganization plan was developed, and on January 14, 1897, the new South Side Elevated Railroad was incorporated. Leslie Carter, president on the new South Side elevated, and a progressive board of directors, were ready to try an untested new electrical system.

Shortly before he was ready to travel to London on the Central London Subway elevator contract, Sprague had been asked by Carter to comment on

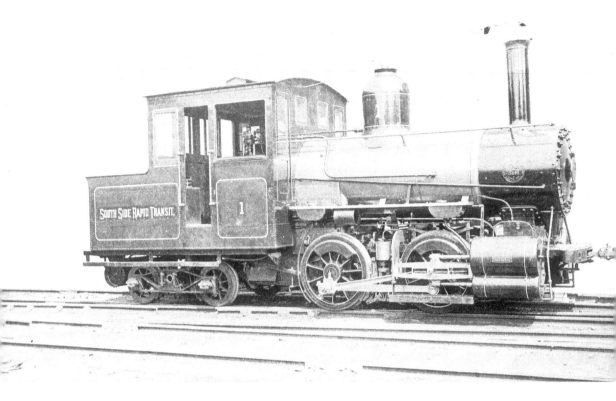

Chicago's first elevated was the Chicago & South Side Rapid Transit Ry., which began operating with steam power in 1892 with Forney-type four-cylinder compound locomotive No. 1, one of 20 ordered by the Baldwin Locomotive Works in 1892. The Intramural Railway's early electrification had worked so well that several other "L" lines were soon converting to electric power. The Metropolitan West Side line was converted to electric power before its construction had even been completed, opening in 1895, and the Lake Street "L" converted to electricity in 1896. The South Side would be the last to electrify, but its multiple-unit control would adopt what would come to be ranked as one of the most important rapid transit advances in history. *Railroad Museum of Pennsylvania.*

the specifications developed by the consulting engineers for electrification of the South Side line, and it would prove an opportunity for Sprague for one of the most important technological advances in electrical railroads.

Busy getting ready for the London project, Sprague was initially reluctant to take on the review of the electrification specifications. But he was persuaded after a visit from Frederick Sargent, a mechanical engineer and old friend who, together with Ayers D. Lundy, had worked with Sprague on the Richmond trolley project and were the mechanical and electrical consultants for the South Side electrification. Formed in 1891, the Sargent and Lundy firm went on to become a major builder of electric power plants.

Sprague's review of the specifications, completed on April 7, 1897, had a number of suggestions for modifications to the electrical plant, but his most significant recommendations concerned the railway's motor equipment. "This is the most serious problem, and my recommendation is radical," wrote Sprague. The two generally accepted types of electrification were then confined either to the use of individual car equipment, which could not be operated in trains, or the locomotive train system, in which a steam or electric locomotive pulled a train of cars. The locomotive system, said Sprague, was the only one which had been considered. The third, said Sprague, is "individual equipment with combination control, which is the system I most strongly advise." Sprague proceeded to enumerate the many arguments for multiple unit control:

> Without further argument, I recommend on your road the absolute abolishment of the locomotive system, tentatively condemned by your engineers by the uncertain and varied suggestions they have made, and the individual equipment of cars with one or more motors, so controlled that one or any number of cars without regard to sequence or position can be coupled together indiscriminately, and from the leading end of any car the train system can be operated.[12]

Sprague's long list of the improved operation that would be possible with the use of the multiple unit system included the use of train lengths from a single car to any number of cars, controlled from either end of the car or train. Train acceleration and deceleration with distributed motive power would be the same regardless of the number of cars, and the use of distributed motive power would permit the reduction of train intervals. The more even distribution of power requirements would lower line losses and reduce the strains in power plant engines and dynamos. And any current failure or an accident to a motorman would bring a train to an automatic train stoppage.

All of this would bring "a very marked increase in the number of passengers carried—the most important net result," concluded Sprague, displaying his usual confidence in his work. "It may be fairly asserted," he said, "that the dividend earning capacity of this single improvement [i.e., multiple-unit control] will be greater than the saving made by the entire remaining cost of the equipment."[13]

Sprague's proposals were met with widespread doubt and ridicule by many in electrical engineering circles. Benjamin E. Sunny, a financier and business executive, recalled the bitter opposition of the two principal electrical manufacturing companies in an enjoyable letter many years later on the occasion of Sprague's 75th birthday celebration:

I was manager of the General Electric Company in Chicago at that time, and we had the business of the Elevated Railroad in motors, etc., for some years, and a contract for equipment had been negotiated which contemplated an addition to the old method, the only one known, of a heavily serviced motor car capable of pulling a number of trailers.

I called on Mr. Hopkins, the general manager, one morning, expecting to receive the signed contract, but found him in rather bad humor. One of the directors had met a fellow in New York with a military title, who had put in an electrically appointed dumb waiter for delivering cocktails to the several floors in a big hotel, which worked perfectly, and he proposed to apply the same scheme to the operation of elevated trains. This seemed to us to be just too funny for anything and we both had a good laugh over it.

Mr. Hopkins telephoned me the next day that the name of the magician was Lieutenant Sprague, and then it wasn't so funny. Your fame in electric railways was widespread, and you were a terrible man to meet in competition. I saw that fine contract in those days of bad business following the '94 panic slipping away, and it made me feel pretty bad.

Then you came to Chicago, and the story was that you had no shop; no organization; no installers—just a tooth brush and an idea, but Ye Gods, what an idea! Most men with an idea like that would not have had the presence of mind to carry even a tooth-brush additional! We lost the contract; you won, and it was a great day for railway transportation, for the application of your multiple-control system has been the greatest boon that has come to that most important public service. Indeed, the wonderful results could not have been secured in any other way.[14]

The Sprague proposal for a multiple unit system was endorsed by Sargent and Lundy, the South Side's consulting engineers, and Sprague supplemented his review report with an April 17 offer to undertake the equipment on the general plan that he had outlined, including electrification of 120 of the South Side's 180 passenger cars with electric lights, heaters, motors, trucks, controls, and braking systems. Other proposals for the work, which included ones from GE, Westinghouse, and Walker Manufacturing Company, using equipment designed by Professor Sidney H. Short, had all recommended the use of locomotive-hauled trains.[15]

After a visit to Chicago, the award of the contract was still undecided. The greatest problem concerned the line's average operating speed. Sprague had guaranteed a 15 mph average speed with the multiple unit system equipment he had specified, while Westinghouse had guaranteed an 18 mph average speed over the line with locomotive-powered trains, using the same total horsepower on a single locomotive car. Sprague was convinced that the Westinghouse speed guarantee could not be met, but South Side president Carter could not immediately be convinced. Finally, Sprague sent his assistant, C. R. McKay, to Chicago loaded with data on accelerations, motor capacities, and such, "with orders to kill that 18-mile schedule beyond resurrection."

McKay set out for Chicago on a Sunday, while Sprague was booked to sail for London on the following Thursday, in late April. Finally, the day before Sprague was leaving for England, Carter wired him that he would accept the schedule which Sprague had guaranteed and would contract with him on the basis of his proposal, but that he wanted a $100,000 bond to guarantee performance. Sprague replied to Carter on April 28, just before he left for London, hoping that the South Side could manage to increase the price by $12,000, allowing for some additional work under the contract that had been asked for. In the end, Sprague got no extra money, and the contract amount remained at $273,000. And without time to establish a bond before his departure, Sprague promised to establish appropriate guarantees upon his return. The final contract was not completed until well after Sprague had left for London, and McKay signed the document for him.[16]

Once again, Frank Sprague had taken on a daunting task. He knew that he wasn't getting enough money for the work. The multiple unit concept was only on paper, and the equipment still had to be designed, built, and tested. To meet the overall schedule he had to begin work on the entire 120 cars, and he had only two-and-a-half months in which he had to have the first six multiple unit cars ready for operation on a mile-long standard gauge track supplied by Sprague (this was provided by GE at its Schenectady plant), and ready for testing by the officers and engineers of the South Side road. Should the test work not be completed on time, or prove unsatisfactory, the contract could be cancelled. Further tests could be called for elsewhere, and the remaining equipments had to be completed by specified dates. As soon as the South Side power plant and road were ready, Sprague was required to make another test of not less than 20 cars under severe conditions over a period of not less than ten days. And once again, should the equipment prove unsatisfactory, the contract could be cancelled, with a waiver of all claims against the "L" company. The equipment would come in batches of 20 cars until the entire project was completed.

Despite the great financial risks of the contract, all of these burdensome conditions were accepted, much as Sprague had taken on the similarly demanding contract for the Richmond streetcar system a decade previously. As he had before, Frank Sprague took on the multiple unit system, confident in his own mind that he could successfully complete it. He would work furiously with his dedicated team to meet the demanding schedule, knowing that the successful completion of the project would open far greater opportunities than any loss he might make on the immediate project.

Off to London on the elevator project, Sprague had to provide most of his instructions on preparing for the trial equipment by cable to his men in New York. A strike by machinists in the shops of the Sprague Co. was a further setback. Problems with settling the contract award for the Central London Railway elevators had delayed Sprague's departure, and it was after the middle of June before he was able to return to New York. Only 30 days were left to complete the construction and preparation of the test cars. Much of the equipment for the 120 cars would come from other suppliers, and Sprague had negotiated conditional contracts under which the suppliers would make available enough equipment for the tests. With only a few weeks to get the equipment ready, Sprague, GE, McGuire, and other suppliers had to work around the clock to get the test equipment ready.

Frank Sprague had long since attained a reputation as a demanding and undaunted leader who would get a job done no matter how difficult the task, and getting the test cars completed and in operation in the limited time available was surely one of the most demanding he had ever faced. Two men at the Schenectady test site years later provided a good indication of the pressure Sprague was under. As engineer Willits H. Sawyer remembered it:

> It was on the berme bank [sic] at Schenectady. The test car was out for your inspection and approval. The mechanism failed to work properly. Your normal restraint broke bounds. There issued from your lips a flow of language that burned, sizzled, and would have withered a longshoreman. As I remember it, the control handle went over into the canal. I certainly had an appreciation that you wanted what you wanted when you wanted it and were going to get it.[17]

"A Master Control Handle did not properly fit," remember Samuel B. Stewart. "The Erie Canal was close at hand. The handle instantly found a resting place in it and for aught I know is still there. 'Get another one boys' was your quick command, and the test continued."[18]

They made it, and on July 16, 1897, two of the six test cars were completed and ready to go into operation on a GE test track in Schenectady that had been made available by GE's railway department manager, William J. Clark, after GE had won the traction motor contract for the South Side electrification. The test cars were standard South Side Elevated Railroad equipment, each of which had been modified on one end with a new McGuire Manufacturing

On July 1, 1897, workers from the Frank Sprague group celebrated the beginning of the first experiment at the track along the berm bank of the Old Erie Canal in Schenectady with cars equipped with the Sprague multiple-unit system.   *Frank Sprague Papers, Manuscripts and Archives Division, The New York Public Library, Astor, Lenox and Tilden Foundations.*

The first pair of cars equipped with the Sprague multiple-unit system made their run alongside the Erie Canal in Schenectady on July 16, 1897. *Frank Sprague Papers, Manuscripts and Archives Division, The New York Public Library, Astor, Lenox and Tilden Foundations.*

Company truck and two GE 57 motors, with a normal capacity of 50 horsepower, and the new Sprague multiple unit control equipment. Sprague had also designed a folding motorman's vestibule which could be used on one side of the open end platform when a car was used as the lead car of a train. When not in use the vestibule could be removed and the platform used for ordinary entrance and exit.

In a little more than another week the work was completed for all six cars, and on the evening of July 25th, a preliminary run was made with all of the test cars. On the following morning, a number of prominent railroad people, as well as officers and engineers form the South Side Elevated were on hand for several days of tests with trains of anywhere from one to six cars, operating at speeds as high as 33 miles per hour. For the first run Sprague's 10-year-old son Desmond operated a six-car train without any difficulty, handling both the switch and the air-brake. Recognizing that it was the 50th anniversary of the late Professor Moses G. Farmer's celebrated electric car test in Dover, New Hampshire, Sprague sent a congratulatory telegram to Farmer's daughter Sarah.

"The leading features of the demonstration at Schenectady," reported *The Electrical Engineer,* "were the wonderful control and smoothness of operation of the system. The starting and stopping were equally prompt, and the train of six cars moved and was operated as if an individual car."[19]

By November 17th a train of five cars were put into service over Chicago's downtown loop and the electrified Metropolitan West Side Elevated Railway

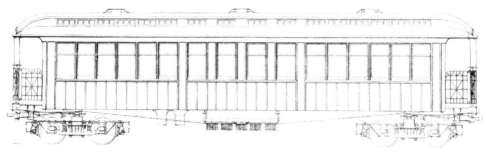

ELEVATION OF STANDARD CAR

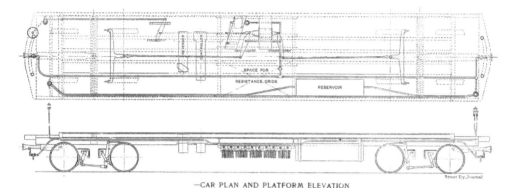

—CAR PLAN AND PLATFORM ELEVATION

In these drawings the arrangement of a standard South Side elevated car is shown with the placement of electrical and braking equipment.   *Middleton Collection.*

Company. Continuing in tests for a number of weeks, the Sprague equipment operated suitably enough for work on the full contract to continue. Francis H. Shepard, the assistant to GE's William J. Clark, worked with Sprague on the development of the South Side's electrical systems, largely supervising the on-site work, and went on to work with Sprague on other multiple unit projects. Shepard would later move to Westinghouse, where he oversaw many of the great electrification projects of the early 1900s.

Other key men in the Sprague work included some from previous projects, such as Hill, Carichoff, and Pat O'Shaughnessy, who had been with Sprague from the beginning, and new men such as Libby, McIver, Pattison, and Campbell.

In a move that *The Electrical Engineer* took as his return to the electric railway field, Sprague established the new Sprague Electric Company at a capitalization of $5 million, combining its work with the Interior Conduit & Insulation Company which had been formed during the previous six or seven years. Thus was the presentation in *The Electrical Engineer:*

Before it had even been tried, Frank Sprague contracted to equip 120 cars with the multiple-unit system for the South Side elevated line in Chicago, provided it could be satisfactorily demonstrated first on 20 test cars. He succeeded, and multiple-unit control ultimately became the standard for rapid transit railways all over the world. Frank is standing on the far left of this test car in 1897 on the South Side line.   *Fred W. Schneider III Collection.*

Particularly significant at this juncture is Mr. Sprague's definite return to the railway field in which, as a pioneer, he made so deep a record, and once again he may be expected to give the art a new date of departure. To-day, even more than when his Richmond road was started, are there difficult problems awaiting solution, and of Mr. Sprague's resolve to deal with them an earnest is now given.[20]

By the spring of 1898 electrification of the South Side road was nearly complete, and the first of the electric cars was operated on April 15th. On April 20th, 20 cars were put into operation, but before the day was out 17 had been taken out of service because of defective rheostats (usually now called "grid

Soon after the cars were operating on the newly electrified South Side "L," a five-car train was posed on the center track above 63rd Street just west of Cottage Grove Avenue.   *Walter R. Keevil Collection.*

resistance"), one car in flames. But Sprague took some satisfaction when the three surviving cars were able to push a stalled steam locomotive around a curve. The problems with the equipment were soon solved, and by July 27th the operation of steam locomotives had been entirely ended by the South Side. The road's initial 120 electrified cars were later supplemented by 30 non-motorized "train line" cars, and another 30 fully equipped electric cars.

An unusual test of the multiple unit control was made by operating two cars together without the benefit of their normal coupling links to demonstrate how closely the multiple unit controls operated. Fred W. Butt, later a New York Central engineer, described one such test:

> One of the tests of operation of the M.U. [multiple unit] system which I feel sure convinced all persons who witnessed it was the running of two cars on the middle track of the Broadway Division with the coupling links and pins removed from between the two cars. With air brake hose coupled, safety chains hooked and electric cable in place, the two cars were considered ready to run. Witnesses were assembled on the two adjacent platforms of the cars where they could note the variation in the space between them as they proceeded along the track. The two cars started at practically the same instant. During acceleration and up to the time the air brakes were applied, there was but a small variation in the width of this space.[21]

The financial results of multiple unit operation were little short of sensational. During the first three months of 1898, when trains were operated entirely with steam power, operating costs for the South Side had consumed

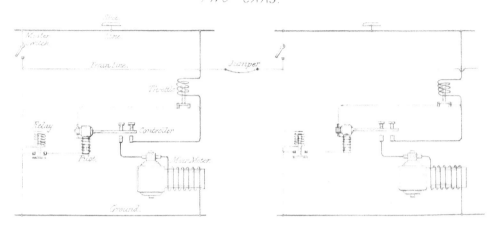

This early drawing, probably made by Frank Sprague himself, shows the elemental circuits of two multiple-unit cars.   *Courtesy of John L. Sprague.*

82 percent of the company's revenue. By the last three months of 1898, when operation had fully gone over to electric service, operating costs dropped to only 57 percent of the road's total revenues, and net earnings for the last half of 1898 were nearly double what they had been in the first six months. Before the South Side's reorganization its stock was selling at $32 per share. After electrification it had climbed to $105 per share.

Electrification got an early test on October 19, 1898, when Chicago's Jubilee Day celebrated the victory in the Spanish-American War. Streetcars were banned from the downtown area, and the elevated lines transported more than twice their usual traffic. The South Side managed to operate an average of 240 cars per hour onto and off of the loop for several hours without difficulty, and the road transported more than 129,000 passengers that day, more than double their normal traffic.[22]

It would be hard to overestimate the importance of Frank Sprague's multiple unit to electric transportation. Sprague had developed some ideas along the lines of distributed electric power as early as his talk to the Society of Arts in Boston in 1885, but he did not develop them until his work on interconnected control of multiple electrical equipment for elevators about 1895, through which any number of elevators could be simultaneously controlled from a single master switch. Early rapid transit electrification had simply followed the pattern used by steam locomotives, which used a single locomotive

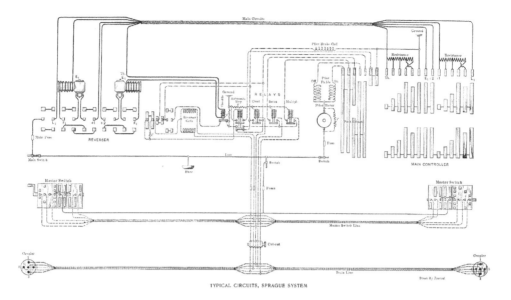

TYPICAL CIRCUITS, SPRAGUE SYSTEM

This more detailed development shows the typical electrical circuits of a Sprague multiple-unit system. *Middleton Collection.*

to pull a train of cars, with an electric locomotive substituting for the steam power. Modifications to this approach followed in 1893 and 1895, when a single electrified passenger car pulled a train of cars, but the idea of a single locomotive pulling a train of cars remained unchanged.

The heart of Sprague's M.U. concept, however, was that it basically changed the way service would be operated. Instead of developing service along the lines of largely fixed-size locomotive-powered trains, interconnected multiple unit power would readily permit the formation of trains, large or small, as needed by traffic, with equipment operating on identical performance standards for trains of any length, and with economy. In writing about the M.U. system in the 1900 issue of the *Transactions of the American Institute of Electrical Engineers,*[23] soon after Chicago's South Side "L" electrification was completed, Sprague described it as

> a semi-automatic system of control which permits of the aggregation of two or more transportation units, each equipped with sufficient power only to fulfill the requirements of that unit, with means at two or more points on the unit for operating it through a secondary control, and a 'train line' for allowing two or more of such units, grouped together without regard to end relation or sequence, to be simultaneously operated from any point in the aggregation. . . . For any given weight to be moved . . . whether it be in one or two cars, there is a certain capacity of motive equipment with which it is

A three-car South Side train rattled along the South Park Way viaduct crossing Garfield Boulevard at 55th Street.   *Middleton Collection.*

best to operate it under fixed conditions, and that is the motor equipment
which should be put on that unit, not something either larger or smaller,
and then when more capacity is required, to simply add another unit of like
character.[24]

The similarity of multiple unit equipment gave like performance to trains,
whatever the combination of cars, since motor equipment was directly proportional to the number of cars. There was a fixed relation between the weight on
the drivers and the total load, whatever the length of train, and a motorman
could be indifferent to whether he was operating a single car or any number
of cars in a train, for its characteristics were the same.

With every car lighted, heated, and braked independently, each car had the
capability of independent movement as needed for inspection, maintenance,

or train operations. The head- and tail-switching needed with locomotives were entirely eliminated, and the flexibility of M.U. equipment made it possible for full or partial trains to easily be reversed at any crossover as needed to meet passenger demand, reducing total dead mileage operated in the system. Wherever a system had main lines and branches, car units for the different destinations could be aggregated on the main line, and quickly separated at the branches, reducing the capacity demands on a busy main line. The ability to readily operate different lengths of trains as required by traffic, while retaining the same performance characteristics, made more frequent service feasible, and more local and express service could be operated.

Multiple unit cars eliminated the heavy loads of locomotives or locomotive cars, and distributed the weight of cars, trucks, and motive equipment throughout a train, reducing the weights on the girders and beams, the strains on elevated structures, and the hammering of cars on rail joints.

Sprague maintained that the distribution of motors and braking equipment throughout the train, providing uniform acceleration and deceleration rates, would safely permit an increased density of operation, while the more flexible operation of the cars would reduce the number of cars required. Multiple unit equipment would also provide greater safety in the event of brake failure or slippery rails by allowing the emergency reversing of traction motors throughout a train.

The operation of multiple unit cars was simple, with every motor car or pair being a transportation unit, and—so far as the motorman was concerned—every aggregation of cars made no change in its operating character. The making up of trains was as simple as coupling an air hose. No main currents were carried car to car, only small currents through reversible jumpers, and the electrical combinations were made automatically.

"Protected by the automatic features, a child of ten years can handle full-sized trains on regular service with less trouble, so far as the electrical apparatus is concerned, and with less instruction than is required for the simplest form of air brake," said Sprague, perhaps mindful of his ten-year-old son Desmond's operation of the test train in Schenectady in 1897.[25]

Multiple unit equipment was easy to inspect. With the train line and main motor circuits being independent on each car, equipment could readily be inspected almost anywhere, and most of the working parts of the motors could be inspected through a trap door in the floor of the car. The use of multiple unit cars also produced economies of operation through reduced wage costs and power costs.[26]

The essentials of multiple unit control were as follows:

- The master controller was located on the platforms at each end of a transportation unit.
- The master controller and train line cables became parts of the permanent wiring of a car. These secondary controlling cables were independent of the main motor circuits, and carried very small currents.
- The jumpers, which were removable sections of the train line, connected the parts of the train line which are permanent to each car, just as air hose couplings connected up a brake line.
- The main controllers, which were composed of relays and a throttle, a pilot motor with automatic limits, a rheostat cylinder, with or without motor grouping switches. The parts were similar to those of hand control, and a reverser with like parts, but independently operated.

In operation, the motorman's master control switch energized certain train line wires, which first made sure the reverser on each car was thrown in the correct direction. Then relays on each car, following the commands of the train line wires, controlled an electric pilot motor advancing the drum controller, which was very similar to an ordinary streetcar "K" controller. The advance was regulated by a relay which sensed the motor current in that car, thus effecting an automatic acceleration control. When the master control switch was returned to the off position, the drums also moved back to their off position.

> The braking system, whether using automatic air or electric, is something like the multiple-unit electric system. There is a train line with means at each end of each transportation unit for simultaneously applying the brakes. When automatic air is used, there is a train and equalizing line, a compressor with an automatic governor, illuminated gauges and a simple form of engineer's valve at the ends of each car for each transportation unit.[27]

Soon after the tests of the South Side "L" cars were made in July 1897, Frank Sprague again returned to his long effort to convince the Manhattan Railway of electrification. A letter to Manhattan president George Gould on December 13, 1897, pointed out the successful work in Chicago and requested an interview, asking "if, after investigation of what has been done, the Manhattan Co. will be inclined to entertain a proposition for the entire electrical equipment of its system, from power house to motors, for a price based in part upon a capitalization of the actual saving accomplished in coal and depreciation, or cost per passenger carried."[28]

But, once again, nothing much happened. Reading in the New York newspapers that the Manhattan was making plans for electrification based upon the locomotive car arrangement, Sprague wrote again on January 12, 1898, this time to discuss the Manhattan's needs, opposing the idea of locomotive cars, and, once again, announced that "we are prepared to install any part or the whole of an individualized equipment, and as against any possible locomotive car plan, on the same rails and with the same dynamos back of us, and using your present cars modified to meet the conditions, operate in any combination of from one to eight or ten cars, and with any combination of rails, under the following guarantees . . ." and went on to enumerate the comparative tests that should be made between an electric locomotive car and a Sprague M.U. car train. Still nothing happened, and he wrote still another letter, this time to John H. Starin, chairman of the Rapid Transit Commission's contact committee on April 1, 1898, to argue that electrification and the higher operating speeds that electric power could provide should be adopted before any rapid transit company should be franchised for extensions. Again, there was no response.

But while the Manhattan Railway still seemed unable to move on electrification, other elevated companies did. Even before the test operation for the South Side "L" began, and when Sprague was still in London settling the Central London Railway elevator contract, his assistant, C. R. McKay, was visited by a delegation from Brooklyn's King County Elevated Railway, who were interested in engaging Sprague as the consulting engineer for electrification of the elevated, operating their road "on as nearly as possible the same general plan [of the South Side 'L']."[29] After visiting Sprague's South Side test installation in Schenectady in July 1897, George Cornell, chief engineer of the Brooklyn Elevated Railroad Company, had publicly stated that it "was the only system which would be accepted on the road."[30]

By early 1898 the Brooklyn Elevated had contracted with the Walker Co. of Cleveland for trucks ands motors, the Pullman Co. for electrical equipment, and the Sprague Electric Co. for the M.U. equipment. Initially the road had contracted for 50 traction motors, enough for 25 cars, with another 100 motors following, bringing a total of 75 motorized cars to the road.[31] Brooklyn began operating its elevated with electric power in 1898, while the Manhattan Railway continued to discuss if—and how—it would upgrade its steam locomotives.

In Chicago, the Metropolitan and Lake Street "L" roads, previously equipped with electric locomotive cars, were converted to multiple unit operation, and the new Northwestern Elevated Railroad was opened as a multiple unit control when it marked its debut in May 1900.[32] Construction of the new Boston Elevated Railway Main Line elevated began in 1899. Planned for electric operation from the beginning, 100 cars for the new elevated were equipped with Sprague multiple unit control when it was opened in 1901.[33]

Finally, in New York even the Manhattan Railway decided to proceed with electrification. In November 1900 a six-car train was tested on the Second Avenue line between 65th and 92nd streets with Sprague M.U. control. The results were convincing, and by May 1901 contracts were awarded for an $18 million project that would be the largest electrification project yet started anywhere. But Frank Sprague lost the big contracts. GE would supply its own version of multiple unit and power supply equipment, while Westinghouse would build the electrical equipment for the Manhattan's enormous new power plant. Some 500 existing cars were converted to multiple unit operation, and 100 new M.U. cars were ordered. The first line to open with electric power was the Second Avenue line, which began operation in January 1902, while the entire Manhattan Line was converted to electric power by April 1903.[34] Early in 1900 construction also began for New York's Interborough Rapid Transit subway, which also would use multiple unit control.

As main line electrification began early in the new century, such early electrified suburban services as those on the Long Island Rail Road (1905), the West Jersey & Seashore (1906), the New Haven (1906), and the New York Central (1907) all employed multiple unit control for their suburban cars.

Overseas, London's pioneer Metropolitan District Railway subway was converted from steam power to multiple unit operation, and multiple unit equipment was sold to London's Metropolitan Railway and the Great Northern & City Railway, while London's pioneer Central London Railway replaced its electric locomotives with multiple unit equipment. Britain's Sprague patents were later sold to the British Thomson-Houston Co. as Sprague-Thomson-Houston. In 1900 French firms joined with Sprague to establish the Société Française Sprague, which supplied multiple unit equipment to the underground railways of Paris and Berlin.

Within just seven years after the first demonstration of Sprague's multiple unit system some 70 roads in the United States and abroad had adopted the system, with more than 3,000 multiple control units in operation with single units varying anywhere from 100 to 3,000 horsepower.[35] And over the next century Frank Sprague's innovative multiple unit control system became the world standard for electric operation of rapid transit or suburban trains.

Working through his Sprague Electric Co., Sprague sold his multiple unit equipment for the Chicago South Side, Brooklyn, Manhattan, and Boston elevated roads, as well as overseas lines. Both GE and Westinghouse quickly entered the multiple unit control field with similar equipment.

Almost immediately, there were disputes between the competitors. GE had developed its own very similar all-electric multiple unit system, while Westinghouse had developed a similar multiple unit system that used a step-by-step pneumatic motor to operate the controller. GE felt that Sprague was infringing upon its motor-control patent and refused to sell traction motors to Sprague, who then bought motors from Westinghouse instead. GE then sued Sprague for infringement of its series-parallel controller patent, which—oddly enough—was originally developed by Sprague and then sold to the predecessor Edison General Electric in the merger with the Sprague Electric Railway & Motor Co. in 1890. Sprague then brought suit against GE for infringement of his multiple unit control system, which he had carefully patented in great detail. Recognizing that Sprague would probably win, GE decided to buy out the Sprague Electric Co., "as in no other way could we get possession of a patent which was absolutely necessary to our business," explained a GE patent department man.[36] This was done by GE with Westinghouse and British Thomson-Houston Co.

Frank Sprague did not want to give up his prospering manufacturing business, but GE had the financial power and considered acquisition of the Sprague firm essential. The sale of the Sprague firm, as spelled out in a 1902 issue of *Electrical World and Engineer,* called for the payment from GE of nearly $2,500,000 to the Sprague firm, including some $1,600,000 for the company's net assets, and for patents, good will, and contracts, while Sprague took up a long-term assignment as a consultant to GE.[37]

Having developed a new electrical industry, as he had done previously for the street railway and the electrical elevator, Frank Sprague sold his manufacturing business to return to the business of electrical invention and consulting.

7

ELECTRIFYING THE
MAIN LINE RAILROADS

Frank Sprague's principal area of interest had shifted from electric traction to the development of the electric elevator with his sale of the Sprague Electric Railway & Motor Co. in 1890, and the subsequent establishment of the Sprague Electric Elevator Co. in 1892. But as we have already noted, Sprague also remained very much interested—and involved—in the development of electric traction. Best known was his continuing (although unsuccessful) effort through most of the 1890s to convince the New York elevated railways to shift from steam power to electrification, and his invention of the multiple unit system for the electrification of Chicago's South Side elevated in 1897–1898. Much less well known was his involvement from 1892 to 1894 in the development of what was the first electric locomotive capable of handling main line railroad traffic.

In 1890 electric railway operation was still confined to street railways. Heavier rapid transit electric railways would soon be used for subway or elevated lines, and a few small electric locomotives for railroad lines were employed for mining or switching lines. But even at this early date, Frank Sprague could clearly see the great potential that electrification might have for main

When this photograph was taken in the mid-1890s, Frank Sprague had barely reached 40, but had already built the world's first successful commercial electric street railway, the electric elevators for what was then the world's tallest building, and the world's largest electric locomotive, and had become the president of the American Institute of Electrical Engineers. *Middleton Collection.*

line railroads. While some early proponents could even see electrics replacing steam power on a wholesale basis, Sprague had a more realistic view of where electrics would have a significant advantage over steam. Recognizing the high cost of building the power plants and distribution systems needed for electrification, Sprague thought they would be limited to those systems which supported an extremely dense traffic, such as the heavy suburban lines operated by the New York Central in New York or the Illinois Central in Chicago, large switching services in major cities, or unusually dense inter-city passenger services such as those between New York and Albany, or between New York and Philadelphia.[1]

THE SPRAGUE ELECTRIC LOCOMOTIVE

Sprague's opportunity to work on main line electrification came about through the long-time interest of railroad financier Henry Villard, who had developed ideas about the use of electricity for railroad operation as early as 1880. The German-born Villard had come to the United States as a young man, working for a number of years as a journalist, and then into finance through his connection with German bondholders of the Oregon & California Railroad. By 1881 Villard had gained control of the Northern Pacific and its presidency and had completed its transcontinental line to the Northwest by 1883.[2]

Through his interest in electrification, Villard had also become an influential backer of the young Thomas A. Edison. In 1880 Edison's first electric railway experiment was built at his Menlo Park, New Jersey, laboratories, using a small narrow gauge locomotive powered from a dynamo, or generator, through the running rails. A year later Villard, now an Edison Electric Light Co. director, wanted to consider the possibility of using electric power for the Northern Pacific railroad. Under an agreement between Villard and Edison, a larger track about 3 miles long was built at Menlo Park and two larger locomotives obtained, and tests continued well into 1882. Villard had agreed to finance the test track if the experiments were successful, and the Edison Light company would take on the electrification of at least 50 miles of line in the wheat-growing fields of the Great Plains. The electrification would have been considered a success if the wheat could be transported at a cost per ton-mile less than that for a steam railroad. Edison completed studies for the electrifications, even planning for generators that would be driven by large windmills in the winds of the Great Plains area.[3] Although Edison's tests had satisfied Villard's contract terms, by this time the Northern Pacific had gone into receivership, Villard was no longer in control of the railroad, and the Edison-Villard Great Plains electrification never materialized.[4]

A decade later, Villard again had an opportunity to develop railroad electrification. With some financial help from his German backers, Villard by 1888 was back on the board of the Northern Pacific, and then in October 1889 became its chairman. By 1890 Villard was organizing the new North American Company, which would absorb and reorganize his railroads and Edison General Electric holdings, and he planned to use North American Co. to develop and test a prototype heavy electric locomotive.

In 1891 Villard had asked Frank Sprague to develop ideas for the application of electric power to his railroads. Sprague had joined with two other electrical engineers to form the consulting partnership of Sprague, Duncan and Hutchinson, Ltd. Dr. Louis Duncan, like Sprague, had graduated from the Naval Academy in the class of 1880, and after several years of navy service he studied for a postgraduate course in physics and electricity at the Johns Hopkins University. Upon completion of his Ph.D. in 1887, Duncan resigned his navy commission to become a professor at Johns Hopkins. During his distinguished career in electrical engineering, Duncan would later serve as president of the American Institute of Electrical Engineers (AIEE) from 1895 to 1897.[5] Dr. Cary T. Hutchinson, who had also completed his doctorate at Johns Hopkins in 1889, worked with Frank Sprague for several years, and then went on to a long electrical engineering career, which included electrification of the pioneer Baltimore & Ohio in 1895, the Great Northern Railroad's first Cascade Mountain electrification in 1906–1908, and the construction of a large hydroelectric plant on the Susquehanna River.

In an 1892 talk given by Frank Sprague as president of the American Institute of Electrical Engineers, Sprague described the extensive testing that was envisioned for the electric locomotive that was then just getting started. A test loop of about 18 miles was planned, and the train would carry not less than 450 tons at 30 miles per hour on a grade of 0.5 percent. Test locomotives would carry their full rated capacity at speeds up to 30 to 60 miles per hour, while operation at speeds of 75 or perhaps even 100 miles per hour were envisioned for the future.[6]

The consultants developed seven or eight different designs for an electric locomotive, finally adopting one that would have the capability of handling the heavy freight in the busy terminal yards of the Chicago & Northern Pacific and Chicago Northwestern system in Chicago. The locomotive would have to have ample power and be controlled as readily and as reliably as a steam locomotive.

The power supply for the locomotive would employ a system of conductors and supporting structure which could rely upon ample and continuous power at all times, at all speeds, on curves, switches, and crossovers, and would have a continuous block signaling system which would not be disrupted by the use of tracks as conductors.

The heavy Sprague Electric Locomotive adopted by the consultants would be (if only briefly) the largest yet built anywhere in the world. The design of the 134,000-pound locomotive followed what came to be known as the "steeple cab" arrangement, with a control cab placed at the center of the locomotive, with an equipment space at each end that sloped downward to give the locomotive crew good visibility from either direction.

Earlier locomotive designs used either a rigid four-wheel arrangement, or a pair of four-wheel trucks mounted on the locomotive platform, while the Sprague design used four traction motors in a 15-foot rigid wheelbase powered through 56-inch diameter wheels. The framing of the locomotive used heavy steel forging with deep pedestals to carry four pairs of massive cast steel boxes which projected inward to form the brackets which carried the motor armatures and their concentric field magnets. The armatures of each of the four motors were rigidly mounted on the axles. A stirrup projected upward from each of the four pairs of boxes to support the elliptic springs, providing an equalizing arrangement supporting the entire superstructure on equalizing springs, while, contrary to what was usually done, the entire weight of the traction motors were un-sprung. To ensure that the four traction motors would operate as a unit, the four axles were connected together through quarter-cranked connecting rods.

The traction motors were compound wound for 800-volt operation, with a pilot lever pneumatic series-parallel control system, which permitted the four motors to be run all in series, then two groups of motors, each in parallel, the groups being in series, then finally all four in parallel. Because of the size of the controlling apparatus the designers decided to provide for power operation with a combination pneumatic system operated from an electrically driven air compressor, probably the first use of a pneumatic control system. The locomotive was rated at 1,000 horsepower, with a maximum speed of about 35 miles per hour, and it was capable of a tractive effort of more than 30,000 pounds. The cab was arranged to carry two trolleys, and Sprague envisioned the use of some sort of rigid overhead conductors.

Supervision of the locomotive's construction was carried out by Sprague, Duncan, and Hutchinson, with manufacture of mechanical parts and assembly done by the Baldwin Locomotive Works, while the armatures were

Working with Dr. Louis Duncan and Dr. Cary T. Hutchinson, Sprague developed this 1,000-horsepower electric locomotive, to be operated in Chicago for companies of Henry Villard. *Smithsonian Institution (Neg. No. Z6807-D), National Museum of American History.*

supplied by the Westinghouse Electric & Manufacturing Co.[7] Tests at the Baldwin Locomotive Works indicated that the locomotive could operate at an efficiency of over 92 percent when operating at 1,000 horsepower, and had the capacity of operating at 1,000 horsepower at a speed of 35 miles per hour over a four-hour period.

Unfortunately, however, by the time the Sprague locomotive had been completed, Henry Villard was no longer able to carry out his plans in Chicago, and the locomotive never demonstrated its capabilities. As the great financial panic of 1893 approached and the Northern Pacific headed into bankruptcy, Villard resigned his positions with both the Northern Pacific and the North American Company, and had largely retired from his businesses by 1893. After the locomotive was finally ready for operation in 1895, it made the initial test runs at the Baldwin plant, and was then stored in the Sprague Electric Elevator Co.'s Watsessing Works in New Jersey, where it was eventually dismantled. The overhead power system envisioned by Sprague was never built.

The 1890s proved to be strenuous—and often difficult—years for Frank Sprague. He had managed to successfully engineer and develop his new electric elevator business to lead that industry to electric power. As we have just noted, from 1891 to 1894 he helped to design and build North America's first heavy electric locomotive. A few years later, beginning in 1897, he developed the new multiple unit control system for electric cars. And throughout the period he continued a strenuous effort to convince the owners of New York's elevated railways that they should electrify.

For some time, Sprague's personal life had not been going well. Intensely committed to his electrical work, he put in his usual long hours to successfully complete whatever project he was engaged in. Sprague's preoccupation with his work led to the dissolution of his first marriage, when his son was 7 years old. Young Desmond saw much of his father and the stress of his work during this period. Years later he recalled one experience with his father during the electrification of the South Side "L":

> The most vivid of my youthful memories were when you took me to Chicago in 1897 on a 48-hour trip that lasted a month. I believe we had rooms at a hotel, but they were seldom used as we slept and ate on the "loop line" most of the time. You had problems; sometimes the motor cars wouldn't go: then someone would mix the jumpers and the respective halves of the trains would start in opposite directions. But to me these were minor annoyances. It was very hot, and the real problem was whether the subsidized Italian fruit vendor would have the bucket of iced lemonade ready when we reached the Van Buren Street spur.[8]

Just a year after the divorce, Frank Sprague learned of his father's accidental death after being hit by a train at a railroad crossing at Rahway, New Jersey. While Frank had not been close to his father after he left his two children with his sister in North Adams, Sprague at least had kept in touch with him, and his father's death left only Frank and Charley alive from the Milford family.

AN AGREEABLE NEW HOME LIFE

Sprague's personal life took a dramatic turn for the better in the spring of 1899, when he was introduced to Harriet Chapman Jones, the daughter of a retired army officer, Captain Henry R. Jones. Almost 20 years younger than Sprague, Harriet had been educated at public schools in New Hartford, Connecticut, and studied music and languages in schools and colleges in the United States, Germany, and Switzerland. The two soon found a strong interest in each other, and they were married in New Hartford, Harriet's birthplace, on October 11, 1899, to begin what by all accounts was a very successful union. Harriet helped create a supportive home for Sprague, and the family grew with the birth of three more children, Robert Chapman in 1900, Julian King in 1903, and Frances Althea in 1906.

The couple established a New York City residence on Riverside Drive. In 1903 Frank acquired a gleaming new Winton automobile and the two enjoyed

The Maples, the summer home of Frank and Harriet Sprague in Sharon, Connecticut. These two photographs were taken after the renovations were made. *(above) Courtesy of John L. Sprague. (below) Frank Sprague Papers, Manuscripts and Archives Division, The New York Public Library, Astor, Lenox and Tilden Foundations.*

drives through the Berkshires. There was a summer-long tour of Europe in 1903, and in 1906 Harriet persuaded Frank to acquire a summer home in Sharon, a small, rural town in the northwest corner of Connecticut. The Sprague home was located along Sharon's South Green, together with other homes occupied by wealthy summer visitors from New York and elsewhere. Officially, it was called The Maples, but the Spragues often referred to it simply as their country home.

Sprague modified the original house and greatly expanded the handsome Colonial Revival structure. Frank was happily involved with the architects, Mann & Mac Neilles, on the alterations to the house, and sent extremely detailed instructions to Edwin Aday, who looked over the house, giving detailed instructions on what should be done or when and where vegetable and flower gardens should be planted,[9] even including in one letter a "check list" for Mr. Aday where he could easily check off everything he was supposed to do.

For almost 20 years the house was a treasured retreat for the Spragues. While the two boys were both born in New York before the Sharon house was acquired, Frances was born there soon after the Spragues had taken possession. New tennis and croquet courts were installed on the 12-acre property, and new flower and vegetable gardens appeared. Sprague became an enthusiastic and accomplished amateur gardener, delighting each summer in his rose garden, and bringing boxes of fresh vegetables or baskets of flowers to his New York associates. Visitors were often entertained. Mrs. Benjamin Harrison, widow of the former president, and her daughter once came for a week-long visit. Weekends in the country often grew to four or five days.

With Harriet's help, Sprague began to take more time to enjoy his family and to expand his social interests, both in Sharon and at home in New York. Aside from his technical work, Sprague maintained a deep interest in art and music, and for many years he and Harriet were regular subscribers to concerts and the opera.

Among the couple's social contacts was a growing circle of associates whose interests went beyond the technical subjects that had for so long engaged Frank's work. Harriet had become a close friend with Clara Clemens, daughter of Samuel L. Clemens (better known under his pen name, Mark Twain), and the Spragues developed a warm association with Clemens during the final years of his life, from the time he moved into his splendid Italianate villa, Stormfield, in Redding, Connecticut, in the summer of 1908 until his death in 1910. The couple had developed associations with a widening circle of well-known writers, artists, and literary people. Together with such writers as

The three Sprague children by his second wife were photographed on the front porch of The Maples circa 1910. At left is Julian King, at center is Robert Chapman, and at right, Frances Althea.   *Courtesy of John L. Sprague.*

Frank and Harriet Sprague posed with their three children on the front porch of The Maples in about 1912. The children are, from left to right, Robert, Frances, and Julian.   *Courtesy of John L. Sprague.*

Lyman Beecher Stowe and Albert Bigelow Paine (later Clemens' biographer and literary executor), the Spragues often gathered for dinner and conversation on summer evenings in Sharon or at Clemens' home in Redding. Sometimes Sprague helped mix the literary subjects with the technical. "In my mind," Paine recalled years later in a letter to Sprague, "is a very clear picture of Mark Twain and you and I, sitting on a couch in his billiard room at Stormfield, the while you explained to us certain curiosities of mathematics."[10]

The Spragues had developed a widening interest in the theater, and developed a long friendship with author and humorist Oliver Herford, who once wrote of Sprague: "He made the world the priceless gifts / Of motors and electric lifts / Without which, it is safe to say / The world would be downstairs today...."

A long-term friend to the Spragues during this period was Brander Matthews, the well-known Columbia University professor of dramatic literature and literary critic. Beginning early in the century, Matthews had been holding Sunday evenings "at home" for playwrights, actors, poets, and critics in New York. The Spragues were introduced to Matthews through Harriet's friendship with his daughter, Edith. It proved to be a warm friendship, and the Spragues soon joined in with Matthews' select circle. For the next several decades Sprague kept his Sunday evenings free in the winter to join the gatherings, which editor and author Clayton Hamilton referred to as "Sunday nights on high Olympus."

The relationship with Brander Matthews provides insight, too, into the loyalty that Sprague so often gave to his friends and family. Sprague's friendship with Matthews continued until both men were in their old age. After the death of Matthews' wife, Harriet took over as the head of the table, and Frank agreed to watch over Matthews' affairs if he were ever in need. Clayton Hamilton, a friend of both men, described their relationship:

> For more than half a century Brander Matthews had known every man of letters, every man of the theatre, every artist who was worth knowing in New York, in London, and in Paris. Yet when he was ultimately incapacitated, the one friend upon whom he relied absolutely to look after him . . . was not an artist, but an engineer, Frank Sprague.[11]

The first several years of Sprague's marriage to Harriet were exceptionally busy ones. He had sold out his electric elevator business several years before, but the flurry of orders for his newly developed multiple unit control system, and his patent battles with General Electric and the Westinghouse Electric & Manufacturing Co. over the multiple unit system, kept him busy through 1902, when the patent disputes were resolved—in Sprague's favor—and Sprague sold out his multiple unit control system to GE. Frank Sprague was once again ready to begin another new project, and it was not long in coming.

ELECTRICS TO GRAND CENTRAL

New York's great Grand Central Station, which carried the New York Central & Hudson River and New Haven trains into Manhattan, had been almost overwhelmed by passenger growth by the end of the nineteenth century. The great train shed of Grand Central Depot (as it was originally called) on 42nd Street was the largest interior space on the continent when it opened in 1871, and it handled more than 130 trains a day. But even these numbers were soon overtaken by the enormous growth of New York City. By 1900 the five boroughs in the newly united city of Greater New York had a population of more than 3.4 million. Suburban population north of Manhattan to the Bronx and Westchester County brought growing numbers of suburban commuters to the trains that operated from Grand Central. Even the massive expansion program of the 1890s, which added additional tracks and enlarged station facilities at Grand Central, could barely cope.[12]

Further expansion of Grand Central's terminal facilities would have required the acquisition of extremely costly real estate in the mid-town area. Compounding the railroad's terminal problems were those created by the smoke and cinders of the hundreds of daily train movements that had become an intolerable nuisance for residents along the railroad's tracks. Even more critical were the operating problems created in the railroad's 2 miles of tunnel or partially covered cut along Park Avenue, where the tracks were often so choked with smoke and steam that it became impossible to read signals.

The solution to Grand Central's terminal problems was the work of an exceptionally able civil engineer, William J. Wilgus, chief engineer and later a vice president of the New York Central. The work involved a series of plans developed over several years that began in 1899, and the key to the solution turned out to be the application of the new field of electric traction. As a key developer of electric street railways and with the recent innovation of the multiple unit control system, Frank Sprague had gained recognition in electrical engineering, and in 1899, he visited William Wilgus. At the meeting Sprague proposed the electrification of New York Central's Yonkers branch, and Wilgus became very interested in the possibilities of electrification. "I remember the day in 1899," Wilgus recalled many years later, "when you first called on me and awakened in me the train of thought which later bore fruit in the change of motive power from steam to electricity in the suburban zone of N.Y. Central and in the resulting creation of the new Grand Central Terminal."[13]

In June 1899 Wilgus completed a plan which proposed the electric operation of suburban trains in the two outside tunnels along Park Avenue and then below East 56th Street through a new tunnel that would carry suburban trains to a below ground loop and terminal beneath the existing surface level terminal tracks. This would be made possible through the use of electric power. Steam-powered trains would continue to use the surface level platforms. Although adopted by New York Central's board of directors, funding was never made available for the 1899 Wilgus plan. Various alternatives were proposed over the next several years as the congestion problems grew worse. A new urgency to solve the Grand Central terminal problems followed a particularly severe collision in the smoke-filled Park Avenue tunnel at 54th Street on January 8, 1902. An inbound New York Central train passed a stop signal, colliding with the rear car of a halted New Haven train, killing 15 commuters. Public outcry following this accident was instrumental in bringing passage by the New York State legislature the following year which prohibited the operation of steam locomotives in the Manhattan Borough after July 1, 1908.

Even before the accident, the New York Central was moving toward electrification. In August 1901 Wilgus had engaged Bion J. Arnold, a pioneering electrical engineer with extensive experience in railroad electrification, to study the practicality of operating the railroad's heavy through traffic with electric power between Grand Central and Mott Haven, a passenger yard just north of the Harlem River crossing. Arnold's report, completed in February 1902, recommended the electrification. Then only a few weeks after the Park Avenue accident, the railroad presented the Wilgus plan of 1899 to the New York State Board of Railroad Commissioners as a means of increasing terminal capacity and minimizing the smoke problems.

Before the end of 1902 the New York Central was firmly committed to an electrification project for its New York terminal operations, even though the details of the Grand Central Terminal work had not yet been settled. In an early 1902 letter to Wilgus, Sprague urged him to consider the electrification of all trains into Grand Central, both suburban and through trains.[14] And another revised Grand Central plan completed by Wilgus in December 1902 did just that, calling for the placement of two levels of electrified terminal tracks below the existing street level, but retaining the old Grand Central Station. However, as he considered various schemes for retaining the old station, Wilgus became convinced that this would be a solution far from ideal. The existing structures and the required new buildings would be in unhappy configuration. Instead, he argued, the old station building and train shed should be removed, and a lofty new building for the station and additional railroad space and accompanying new revenue-producing buildings be constructed, making useful the air rights that would become available by eliminating the smoke and gases from the railroad yards. By March 1903 Wilgus had laid out a plan which would incorporate all of the essential elements of the new Grand Central Terminal. The terminal would incorporate a 57-track, double-level terminal, all of it below ground, and a new terminal and office buildings, together with other revenue-producing buildings constructed upon the "air rights" above the terminal tracks. An unprecedented electrification of the New York Central would make it all possible.[15]

Until his departure from the railroad early in 1907, Wilgus would head the enormous task of planning and building the new Grand Central Terminal and the New York Central's electrification. Vice president and chief engineer Wilgus would be the chairman of the railroad's advisory Electric Traction Commission, which would oversee the plans and methods developed for the railroad. Wilgus appointed the outside members of the commission on De-

New York's great Grand Central Terminal, made possible by the electrification of the New York Central railroad. This photograph, from April 1937, looks northward from a point just south of 42nd Street and Park Avenue.   *David V. Hyde, from Penn Central Company.*

cember 15, 1902, with the first meeting to come the very next day.[16] Frank Sprague would be a key member of the commission, which also included two other consulting electrical engineers, Bion Arnold and George Gibbs. Another experienced railroad electrification man, Gibbs had graduated from engineering at the Stevens Institute of Technology in Hoboken, New Jersey, and had previously carried out several important electrification projects as chief engineer for the Westinghouse companies in Europe. He would later work as a consultant for the new Interborough Rapid Transit subway and for Pennsylvania Station. Other members of the commission included Edwin B. Katté, New York Central's electrical engineer and also the secretary to the commission, and the railroad's superintendent of motive power, Arthur M. Waitt, who shortly afterward was replaced by John F. Deems. For the next four years members of the commission met weekly to review and recommend the plans for the New York Central's railroad electrification planning and development, and to review the plans and specifications to carry out the work.

One of the first to be made was a determination of the planned scope of the electrification. While the New York legislature only ruled out steam power operation in Manhattan Borough, south of the Harlem River, an interchange of steam and electric power at that point was not feasible. Possible changes from electric to steam power were Mott Haven, Highbridge, and Kingsbridge in the Bronx, but the commission was quickly persuaded that a much greater scope of electrification was desirable. A transition of motive power somewhere in the railroad's suburban zone would be both inefficient and highly unpopular for passengers. According to Sprague:

> In determining the extent of equipment, it must be borne in mind that the successful use of electricity in a part of the city would create a public demand which would soon be crystallized into law that such use should be extended to the city limits; and also that there would be demand on the part of those who reside within what may be fairly termed the suburban radius to be carried their full trip electrically without change of cars.[17]

There was general agreement with the desirable extent of suburban traffic, and it was determined that electrification would be extended to Croton-on-Hudson on the Hudson Division and to North White Plains on the Harlem Division, which were 33 miles and 24 miles, respectively, from Grand Central. These two locations would be beyond the ordinary limits for travel of an hour's maximum, and both had adequate space to construct the necessary interchange facilities.

Also needed was an early decision on the type of electric power that would be used. Virtually everything that had been electrified so far in the United States—in street railways, rapid transit lines, and what little main line electrification had been done—were electrified at 600 to 700 volt DC, which by this time was a well-proven system. But the newer development of high voltage, single phase AC operation offered great promise in both the first cost and operating cost savings by greatly reducing the required frequency of sub-stations and substantially reducing the necessary conductor size.

The members of the commission to a man were in favor of using the more proven DC system. Both Frank Sprague and William Wilgus had very strong opinions in stating their preference for DC electrification. Sprague was particularly troubled by the required date of the changeover from steam to electric power, as well as the requirement to provide reassurance that the exacting service would be ready to operate in time.

> But hopeful and confident as I, as well as other engineers are of the practical outcome of this development, and promising as many of the experimental trials in different parts of the world [are], we can not avoid the existing legal and commercial necessities of operation, nor afford to jeopardize absolute success by any consideration of what the future may hold for other and different problems, nor allow a moderate saving in first or cost of operation challenge the unknown difficulties invariably attending all new developments.[18]

Considering the bids that had already been received for the electrical work, Sprague noted that a wide variety of systems that were proposed in operating practice and combinations of equipment had been offered for AC work, which made it difficult to select which AC proposal would be used. In contrast, the known possibilities of DC equipment were based on the result of 20 years' development of every variety and condition. Sprague had other concerns. The use of untested new AC equipment would probably require early replacement as more perfect designs were developed. The overhead lines that would be required for AC electrification would present difficult problems, and probably the Park Avenue Tunnel would require deepening to carry the overhead wires. Prospective interchange of traffic and equipment with the elevated and underground trains electrified at 600 volts DC would become impractical.

While Sprague and the other members of the commission took the cautious approach with the proven DC equipment, the New York, New Haven & Hartford, which would also operate over New York Central tracks to Grand Central, took the opposite choice. From its junction with the New York Central at Woodlawn, in the Bronx, to Stamford, Connecticut, a distance of 21 miles, the New Haven installed a new and untried Westinghouse single phase AC electrification. Although both AC and DC systems had been considered by the New Haven, its choice of the AC system undoubtedly was based upon its greater suitability for an already planned long distance extension to New Haven or even, eventually, to Boston. Both of the new electrifications, it turned out, were completed in time to meet New York's required termination date for steam operation south of the Harlem River, and both systems worked well. But for as long as he was involved in any railroad electrification project, Frank Sprague continued to favor the use of DC electrification.

Still another early decision required the determination of the current carrying arrangement between the tracks and trains. Having settled on low voltage DC, the New York Central would have to operate with a heavy third rail in order to carry the large current that would be required. In the early years of railroad electrification in New York, it was envisioned that extensive interchange of electric equipment would be developed between both the electrified trunk lines and the elevated or subway rapid transit routes. Early in 1903 the several railroad and rapid transit companies had developed standards for the placement of third rail installations and dimensions of passenger cars that could operate in interchange service. With extensive construction of rapid transit third rail lines already complete, this proposed standard adopted the over-running third rail system already in use on the elevated lines. This arrangement mounted the heavy steel third rail on one side of a track and 3½ inches above the level of the top of the track, with an iron collector shoe running on the top of the third rail.

While the Electric Traction Commission had tentatively adopted the third rail standard on May 3, 1903, vice president Wilgus began to have some second thoughts about the wisdom of using the over-running third rail. The exposed 660-volt third rail, Wilgus felt, presented a serious hazard to railroad workers and the general public alike, and the top of the rail exposed it to the hazards of sleet, ice, and snow that might often disrupt operations. Frank Sprague was asked to work with Wilgus to design an under-running third rail that would solve these problems.

The design developed by Wilgus and Sprague used a cast iron bracket bolted into an extended tie and carried an under-contact third rail clasped in insulators hung by hook bolts from the top of the bracket. The flexible insulating material surrounded the top and sides of the third rail, protecting it from corrosion, accidental contact, and from sleet, snow, and spray. A section of the new under-running design worked well on a test section of electrified line installed in Schenectady. The previously approved third rail standard was discarded and the new Wilgus-Sprague design was adopted for the Grand Central electrification.

While the potential advantages of equipment interchange of third rail equipment was lost, the Wilgus-Sprague under-running design proved to be far superior to the over-running design used by the Long Island Rail Road and the rapid transit lines in New York. In a letter to Frank Sprague several years later, for example, Wilgus compared the Long Island's operation in a heavy 1913 snowstorm with that of the Central. In a heavy snow the Long Island's

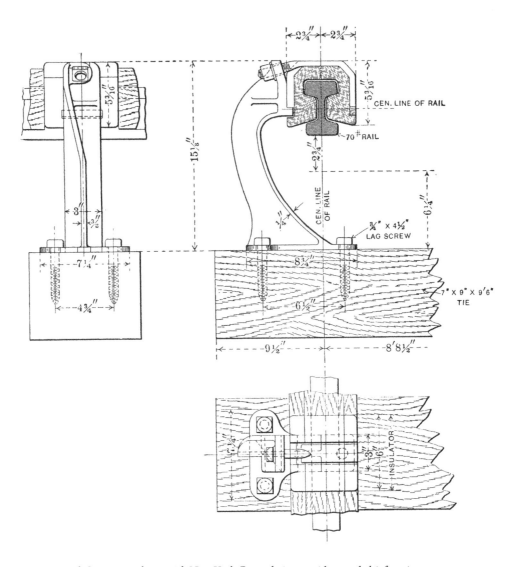

In 1903 Frank Sprague took on, with New York Central vice president and chief engineer William J. Wilgus, the task of designing a new type of protected third rail. The design provided an under-running third rail surrounded on the top and each side by a flexible insulating material, which protected the third rail from corrosion, accidental contact, sleet, snow, and spray. This proved to be far superior to the usual over-running design. Wilgus and Sprague patented the design and sold it to a number of railroads. *Middleton Collection.*

third rails were completely blocked up on the Far Rockaway branch. "Long Island Road Has to Give it Up," reported the *New York Times,* while the Central's under-running design operated perfectly throughout the storm. Wilgus wrote:

> It would seem to me extremely wise on the part of the large manufacturing companies to endeavor to bring about the adoption of working conductors

that will not cause a suspension of traffic during snowstorms, to the inconvenience of the traveling public, the expenses of the railroad company, and, above all, from their standpoint, the damaging of the fair reputation of the electrification of railroads.[19]

Although the new Grand Central Terminal itself would not be complete until early in 1913, electric operation began more than six years earlier. Initial orders for locomotives, substations, and other equipment were placed in the fall of 1903, and a prototype of the electric locomotives began tests only a year later. Design of the electric locomotives was the work of General Electric's Asa F. Batchelder, but the first 35 electric locomotives, built by General Electric and American Locomotive Company, followed some of the same basic ideas that Sprague, Duncan, and Hutchinson had developed a decade earlier for the still-born locomotive for the North American Co. Although much larger, the 225,000-pound, 1,700-horsepower locomotives used a similar steeple cab arrangement that provided good visibility in either direction, and employed a rigid frame that carried four traction motors in 44-inch wheels in a similar "gearless" arrangement that attached the armatures directly to the axles. The field magnets were carried on a horizontal bi-polar arrangement as an integral part of the locomotive frame and were mounted above the suspension springs. Originally, the locomotives had a single-axle guiding truck at each end. Early in 1907 one of the new electric locomotives derailed and overturned in the Bronx at great loss of life. Although no conclusive determination on the cause of the accident was ever reached, these locomotives were redesigned to provide a two-axle guiding truck at each end to improve tracking, and some of these highly successful locomotives remained in service for almost 80 years. An initial fleet of 180 new multiple unit cars for suburban services used such familiar Sprague equipment as "wheelbarrow" traction motor mountings, and multiple unit controls, now sold as Sprague–General Electric after the Sprague patents were sold to GE and Westinghouse in 1903.[20]

Deliveries of equipment and the construction of the power system and generating plants had advanced far enough to pull the first electric train out of Grand Central Station on September 30, 1906, but it would be another seven years before the full planned electrification reached all the way to Croton-on-Hudson. When all of the planned work and its extensions had been completed, the New York Central's pioneer electrification had reached some 500 track-miles of electrification. By the late 1920s Grand Central electric trains averaged about 475 daily trains, with a peak of as many as 800 daily trains, a daily average of about 134,000 passengers, and a peak of more than 166,000.[21]

Designed by Asa F. Batchelder of General Electric, the New York Central's new T-class
electric locomotives departed materially from most other designs. But one important similarity was the use of a bipolar gearless arrangement for its four traction motors that was almost
identical to the one built for the Sprague, Duncan, and Hutchinson locomotive in 1895. Each
of the New York Central locomotives placed the four 550-horsepower electric motors with
the armatures carried directly on the driving axles, while the field magnets and pole pieces
were carried on the locomotive frame, eliminating any need for a geared connection between
the armature and driving wheels. No. 6000, a prototype of the 95-ton locomotive, was tested
on a five-car train at Wyatts Crossing, New York, near Schenectady, on November 12, 1904.
The remainder of a 35-unit order was delivered two years later.   *Industrial Photo Service.*

Guided by Electric Traction Commission engineers William Wilgus, Frank
Sprague, George Gibbs, Bion Arnold, John Deems, and Arthur Waitt, the New
York Central had completed with great success what was by far the greatest
railroad electrification yet built anywhere in the world.

In addition to the use of the innovative Wilgus-Sprague third rail design
on the New York Central's New York project, the W-S Standard Under-Running Third Rail, as it was called, was adopted at a number of other locations as
well. The New York Central used it for the Michigan Central's electrification
through the Detroit River Tunnel and its West Shore interurban line between
Syracuse and Utica, New York. Wilgus and Sprague patented the design in
1906–1908, and the two formed a new company, Standard Third Rail Company, in 1911. Rights to use the W-S design were sold to such new third rail
electrifications as the Philadelphia Rapid Transit, Central California Traction
Company, and the Belgian State Railways. The Central California line, an
interurban, successfully used the W-S third rail electrification for a 1,200-volt
DC installation. But by this time new railroad electrifications were largely

being built for the highly successful high voltage, single phase AC systems, or the newly developed high voltage DC systems, and there was little market for new third rail systems, and Wilgus and Sprague closed down their Standard Third Rail Company in 1918.

POWER TRANSMISSION AND DISTRIBUTION

From the time when Frank Sprague first began to consider the possibilities of railroad electrification, he was interested in how electric power could best be transmitted and distributed. Probably his earliest serious look at electrification came from his visits to London's steam-operated Metropolitan District Railway underground railway in 1882. Sprague quickly became convinced that electric operation would make the subway journey a much more pleasant one, and he developed conceptual ideas of how they would operate. Electric trains would operate in two planes, making electrical contact with them at upper and lower planes. Running tracks on the lower plane would make contact through running wheels, while the upper plane would use a spring operated supported device held against an overhead rail following the center line of all tracks and switches. Some of these ideas were later used in developing Sprague's early proposals for electric rapid transit.

As Sprague began to develop electric street railways in 1886, his systems used an overhead power wire supplied from a spring-mounted trolley pole to collect current. This became almost a universal way of delivering power to street railways, using direct current power generated and distributed at about 400 to 500 volts. The need to move power at these relatively low voltages severely limited the distance it could be carried because of the high power losses with a large current requirement. Sprague's design of the power plant for the Richmond electrification was one of the largest yet built anywhere, and his design of a centrally located plant with power lines no longer than 3½ miles permitted an efficient operation. In the early years of electrification, before the availability of alternating power, Sprague recommended the use of larger generators, operated at the highest possible voltage, as the most efficient method of power distribution. In a November 12, 1888, lecture to The Franklin Institute, "The Transmission of Power by Electricity," Sprague presented an extensive discussion of the transmission of electrical power, emphasizing the development of large generators operated at the highest feasible voltage.[22]

By the mid-1880s George Westinghouse had acquired rights to use the patents developed by Lucien Gaulard of France and John D. Gibbs of England for a jointly developed system of alternating current distribution. Assembling an impressive group of electrical engineers, Westinghouse Electric & Manu-facturing Co. soon rapidly developed AC electrification as a formidable rival of the Edison firm, which was firmly committed to DC electrification, and the fierce "battle of the currents" raged between the Westinghouse and GE firms until the two firms had licensed each other to manufacture almost every type of electric equipment. Guido Pantaleoni, one of the early Westinghouse elec-trical engineers, remembered when Frank Sprague had been tasked to study the two rival technologies while he was a consultant for Edison GE after his sale of the Sprague Electric Railway & Motor Co. in 1889–1890. "Sprague's re-port was an exhaustive one, and a remarkable one considering the atmosphere of hate then existing"; recalled Pantaleoni, "his analysis of the very crude state of the art as it then was and the immense possibilities of the use of AC, could not have been bettered today."[23]

In an extensive paper, "Coming Development of Electric Railways," given by Frank Sprague at his address as president of the American Institute of Electrical Engineers in Chicago in 1892, he again touched upon the prob-lems of electrical power distribution. Noting that the then-booming electric street railway industry was almost exclusively wired with overhead power poles, Sprague still believed that a better solution could be found. On streets also occupied by elevated railways, for example, the elevated structures them-selves could become a practically rigid support structure for power supply to the surface-level street railways. Elsewhere, the cities should allow only the best and least intrusive street railway wiring. The only overhead wire should be a contact wire, with all main conductors and feeders placed in under-ground conduits. For even this, maintained Sprague, "the cable will hold its own [only] until an electric conduit or surface contact system shall be proven satisfactory."

For a proposed electrification of heavy switching lines around Chicago, Sprague believed that only a substantial overhead power system would work, and for a proposed electric express between New York and Philadelphia in 1891 he had suggested electrification with an overhead rod above the cars, with a return circuit through the rails.[24]

The rapid development of electrification technology soon brought exten-sive new railroad electrifications. While early electrifications had been con-fined to direct current systems, the growth in alternating current technology brought a number of new systems into railroad use soon after the beginning of the twentieth century. Most American AC electrifications were single-phase systems, but in Europe there were also advocates of poly-phase systems. In an extended paper presented to the 219th meeting of the American Institute of Electrical Engineers in New York on May 21, 1907, "Some Facts and Problems Bearing on Electric Trunk-Line Operation," Frank Sprague outlined the status of railroad electrification. While he saw great advantages in the use of AC power for power transmission, Sprague had a clear preference for the use of direct current motors as the traction motors, and many of the participants at the May meeting had strong disagreement with many of the conclusions reached by Sprague.[25]

In the New York Central electrification, for which Sprague was a member of the Electric Traction Commission (1902–1906), he had strongly approved the use of the proven DC power for the system. AC power was used to link the power plant with sub-stations on 11,000-volt AC high-tension lines, and then converted to 660-volt low voltage DC with synchronous converters from the Wilgus-Sprague under-contact third rail system.[26] Sprague much preferred third rail power distribution to any type of overhead system, and in his later project for the Southern Pacific he would propose the use of 1,200-volt or even 1,800-volt DC third rail power supply. One interurban company, the Michigan Railway, even installed a 2,400-volt DC power system supplied from a third rail. But third rail power supplied at such high voltages posed serious hazards, and almost all high-voltage DC systems went to overhead wire systems.

Sprague continued to study the development of railroad electrification, and he maintained his preference for direct current operation even in a 1932 study of electric traction. While two major railroads were electrified with 25-cycle (or Hertz), single phase AC systems, he noted, several major new electrifications, including the Butte, Anaconda & Pacific, the Chicago, Milwaukee, St. Paul & Pacific, the Delaware, Lackawanna & Western, and the Illinois Central, were all employing a high-voltage DC power supply at anywhere from 1,500 to 3,000 volts.[27]

ELECTRIFICATION OVER THE SIERRA NEVADA

Close upon completion of Frank Sprague's work on the New York Central electrification in New York came an even more demanding challenge for electrification of the route over the rugged Sierra Nevada mountains of Southern Pacific's transcontinental railroad.

In 1900 Edward H. Harriman had taken control of the Southern Pacific which he would operate in tandem with the Union Pacific. Harriman had already begun major improvements to the Union Pacific, and soon started

similar work for the badly run-down Southern Pacific. Over the next eight years Harriman would spend $247 million to make SP the equal of the Union Pacific, cutting its maximum grades, reducing its curvature, building a new entry into downtown San Francisco, and constructing the massive Lucin cut-off across the Great Salt Lake.[28]

Harriman soon became convinced that electrification would have substantial advantages over steam power. Late in 1905 the Southern Pacific com-

Since the completion of the transcontinental railway in 1879, the crossing of the Sierra Madre Mountains had been a very difficult one, the locomotives fighting against steep grades, sharp curves, extremely heavy snow falls, and fierce storms. These two woodcuts, which appeared in the February 10, 1872, issue of *Harper's Weekly,* gave a good idea of the problems encountered. Frank Sprague designed an electrification over these mountains that would have been by far the most difficult electrification ever encountered. Sprague was confident he could succeed, but Southern Pacific management decided not to go ahead and instead began to acquire the railroad's celebrated cab-forward articulated steam locomotives.

pleted cost studies and engineering details for the reconstruction, expansion, and electrification of the railroad's extensive suburban service in the East Bay area across San Francisco Bay. By the following August it was announced that Southern Pacific president Harriman had authorized the suburban electrification project, which by the time it was complete in 1912 would have cost $10.6 million.[29] Early in 1907 he announced that the SP was appropriating another $5 million for the standard gauge conversion and electrification of the narrow gauge line from Oakland to Santa Cruz, plus acquiring an extension to Salinas, that would create a 128-mile electrified line that Harriman envisioned could eventually be part of an electrified line extending all the way to Los Angeles.[30] Before the year was out Southern Pacific had still other electrification plans to consider, including a project that would extend all the way from San Francisco to Sacramento; a feasibility study made by General Electric at Harriman's request had estimated that the SP could cut its operating costs over the line by 38 percent with electrics.[31]

But by far the biggest news on prospective SP electrification came in August 1907 when the railroad announced its plans to study the feasibility of electrifying the transcontinental Sierra Nevada route to increase capacity of the heavily used line. Electrification of the 134-mile Rocklin, California, and Sparks, Nevada, line was what the *Railroad Gazette* called "perhaps the most difficult and important installation which has so far been seriously considered."[32] The line crossed the Sierra Nevada Mountains at an elevation of 7,018 feet with heavy traffic on a single track line, and with sharp curves and more than 31 miles of tunnels and snow sheds. From Rocklin to the summit of the eastbound line it climbed through 6,768 feet, with a maximum grade of 116 feet per mile, or 2.2 percent grade. Westbound trains climbed 1,198 feet between Truckee and the summit, with a maximum grade of 105 feet per mile, or 2 percent grade. During the winter the snow often accumulated to a depth of 15 to 20 feet in exposed places.

Southern Pacific had considered other possible alternatives to increasing capacity, including a new second line on a new alignment, or the construction of a long tunnel under the mountains. If electrification was used, they had to consider the problems of installation and maintenance of transmission lines under heavy snows and violent storms, the ability to provide power supply during wide variations in load, the danger of fires in the snow sheds during the summer or from short circuits from melting snows during the spring, the cost of overhead line construction and danger to trainmen in tunnels and snow sheds, or the alternate problems of heavy snows if third rail electrification was used.[33]

Julius Kruttschnitt, director of maintenance and operation of the Harriman lines, appointed Frank Sprague as a consulting electrical engineer, together with Allan H. Babcock, the SP's electrical engineer, to complete the electrification study and to prepare detailed information and a general plan of electrification. The two would also serve as members of a Committee on Electrification of the Sierra Nevada Grade, which also included John D. Isaacs, consulting engineer; William Hood, chief engineer; and W. E. Calvin, vice president and general manager. The committee would report their findings and recommendations to Kruttschnitt. In the event the installation went ahead, Southern Pacific staff would carry out the details of plans, specifications, purchases, and installation, while Sprague would continue as consulting electrical engineer during the work.[34]

Work on the Sierra Nevada study had actually begun some three years earlier when Allen Babcock, later with Frank Sprague, began to gather material on possible electrification. The two men worked extensively with early proposals and information provided by General Electric and the Westinghouse Electric & Manufacturing Co. In 1906 both Westinghouse and GE had submitted studies for electrification. The Westinghouse report, submitted in May 1906, proposed the use of locomotives powered by single phase, 25 Hertz AC power, with separate locomotives required for high-speed passenger service and low-speed freight service. General Electric proposed either 1,200 volt DC or single phase Hertz power.

Anticipating that water power would be used for the electrification, the Southern Pacific had begun getting proposals about water availability along the north and south forks of the Yuba River, and obtained permits for building water-powered generating plants along the Rubicon River. By mid-1909 the railroad had applied to the government forest preserve for permits to build plants in Sierra County, Nevada.[35]

In developing the report on the Sierra Nevada electrification, Frank Sprague had to consider not only the present operation but a variety of other improvements contemplated by the Southern Pacific, some already begun while others were still under consideration. Already under construction was a new 35-mile second track on the west slope between Roseville and Colfax, which would reduce the eastbound maximum grade from 2.01 to 1.5 percent, and a 37-mile relocation on the east slope between Truckee and Sparks. A series of proposed projects across Donner Pass would reduce the ruling grade to 1.5 percent with a new 5.2-mile tunnel west of Donner Lake and some 30 miles of new connecting line on either side of the tunnel.

Southern Pacific's electrification committee submitted Frank Sprague's final report to Julius Kruttschnitt on January 23, 1909, recommending the adoption of electric operation over the Sierra Nevada Mountains from Roseville to Sparks, including the development of a hydro-electric power station. The committee recommended that locomotive specifications should be defined only in general terms, without specifying any particular favored electric system. It requested bids for not less than 44 locomotives, and issued the announcement that the Southern Pacific had definitely decided to adopt electric operation for the Sierra Nevada district.

While Frank Sprague had never cared for overhead wiring for railroad electrification, preferring third rail, locomotives for the Sierra Nevada electrification could be designed for DC operation at 1,250 and 2,500 volts, single-

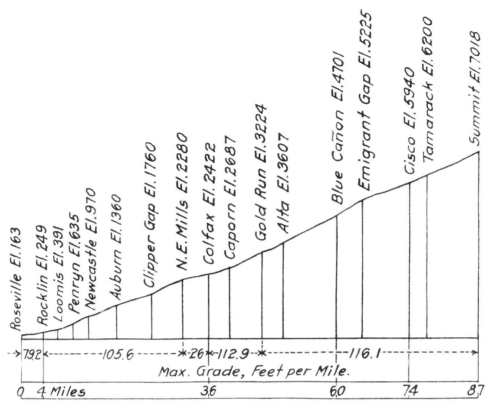

Profile of Road; Roseville to Summit.

This drawing from the January 4, 1910, issue of *Railway Age Gazette* shows the difficult line along the western slope of the Sierra Nevada Mountains that would have been required by Southern Pacific electrification. The road profile is frequently as steep as 2.1 percent, and a train must climb almost a mile and a third between Roseville and Summit.

phase AC operation not exceeding 11,000 volts, and poly-phase operation not exceeding 6,000 volts. Only following a decision on the operating power system would the details of power houses, substations, transmission lines, and the working conductors be developed.

The general requirements for the locomotives specified a continuous drawbar pull on 1.7 percent grade, exclusive of engine requirement, of 20,000 pounds, at both 15 and 30 miles per hour, and a maximum drawbar pull, exclusive of engine requirements, of 40,000 pounds. Each freight locomotive would have a capacity to haul a trailing load of 500 tons, exclusive of caboose, at an average speed of 15 miles per hour, while passenger locomotives would be able to pull a train with a trailing load of 350 tons at 20 miles per hour. Operation of locomotives in two units or more was anticipated, and requirements specified that the locomotives be equipped for multiple unit control, while electric

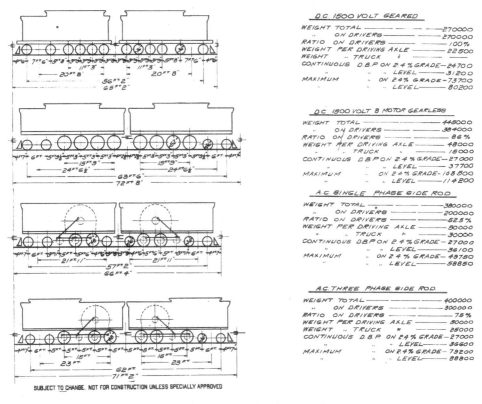

D.C. 1500 VOLT GEARED

WEIGHT TOTAL	270000
" ON DRIVERS	270000
RATIO ON DRIVERS	100%
WEIGHT PER DRIVING AXLE	22500
WEIGHT " TRUCK	—
CONTINUOUS D.B.P ON 2.4% GRADE	24700
" " LEVEL	31200
MAXIMUM " ON 2.4% GRADE	73700
" " LEVEL	80200

D.C. 1500 VOLT 8 MOTOR GEARLESS

WEIGHT TOTAL	445000
" ON DRIVERS	384000
RATIO ON DRIVERS	86%
WEIGHT PER DRIVING AXLE	48000
" " TRUCK	15000
CONTINUOUS D.B.P ON 2.4% GRADE	27000
" " LEVEL	37700
MAXIMUM " ON 2.4% GRADE	108500
" " LEVEL	114200

A.C. SINGLE PHASE SIDE ROD

WEIGHT TOTAL	380000
" ON DRIVERS	200000
RATIO ON DRIVERS	52.5%
WEIGHT PER DRIVING AXLE	50000
" TRUCK	30000
CONTINUOUS D.B.P ON 2.4% GRADE	27000
" LEVEL	36100
MAXIMUM " ON 2.4% GRADE	49750
" " LEVEL	58850

A.C. THREE PHASE SIDE ROD

WEIGHT TOTAL	400000
" ON DRIVERS	300000
RATIO ON DRIVERS	75%
WEIGHT PER DRIVING AXLE	50000
WEIGHT " TRUCK	25000
CONTINUOUS D.B.P ON 2.4% GRADE	27000
" " LEVEL	36600
MAXIMUM " ON 2.4% GRADE	79200
" " LEVEL	88800

SUBJECT TO CHANGE. NOT FOR CONSTRUCTION UNLESS SPECIALLY APPROVED

215394 TABULATION OF LOCOMOTIVE PROPOSITIONS FOR
SOUTHERN PACIFIC COMPANY.

These drawings developed by General Electric proposed four different possibilities for the electrification of the Southern Pacific, and provided outline specifications for each type. *Frank Sprague Papers, Manuscripts and Archives Division, The New York Public Library, Astor, Lenox and Tilden Foundations.*

braking was required both for energy recovery from down grade trains and assistance to air braking on mountain grades.

Sprague put the estimated cost for electrification at almost $6.5 million, including the installation of a water power plant, for operation of the Sierra Nevada line on its existing alignment operating at the 1906 tonnage level. Electrification costs went up to almost $7.2 million based upon a projected tonnage level almost double that of the 1906 level and the completion of all of the SP's projected line improvements, including the 5-mile-long tunnel near the summit. Projected costs of electric operation at the 1906 level of traffic over the existing track would reduce annual operating costs by more than $400,000, more than 25 percent below steam operating costs, which exceeded $1.5 million. While the proposed new tunnel and other line improvements would reduce annual steam power costs to less than $1.4 million for even the

anticipated increased tonnage levels, the annual savings from electric operation of almost $640,000 would reduce operating costs by more than 50 percent
from even the new steam costs.

Still other advantages of electrification cited by Frank Sprague included
higher average operating speed, elimination of stops for fueling and watering
of steam locomotives, and the high overload capacity that would be available
from the electrics.

Julius Kruttschnitt approved the committee's recommendations on April
26, 1909, and asked Sprague to prepare the necessary specifications, with the
electrification plan modified to provide only an initial 15 electric locomotives,
sufficient to power passenger trains, with freight equipment to follow later. Requests for sealed proposals were sent out to General Electric, Westinghouse,
and European supplier Ganz and Company, with proposals to be received on
January 1, 1910, later extended to March 1, and then to May 16, 1910.

But even while the bids were being requested, Edward H. Harriman, SP
electrification's chief proponent, died at his country home in Arden, New
Jersey, on September 9, 1909. And while Southern Pacific's electrification
committee and procurement staff were moving ahead to get proposals for the
electric locomotives, Julius Kruttschnitt and other senior operating staff were
now looking at the alternative provided by the railroad's newest steam locomotives, its first 2-8-8-2 articulated compound Mallet type units. Two of the
new Mallet's—Numbers 4000 and 4001—were delivered from the Baldwin
Locomotive Works in 1909, and soon went into test service against the heavy
2-8-0 locomotives that had previously handled Sierra Nevada freight trains. As
early as July 1909 Kruttschnitt told an interviewer that the new Mallet locomotives had been used so successfully on the mountainous Sacramento division
that the electrification "cannot be considered a probability."[36]

In test runs comparing the two locomotives in June and August 1909, the
213-ton Mallets showed they could easily haul twice the tonnage of the Con-
solidations, with a significant improvement in efficiency, reported the *Railway
Age Gazette*.[37] Kruttschnitt was even more positive in his assessment of the
performance of the new compound Mallets.

> Eastern critics may be inclined to the opinion that we are dallying with
> this matter. We have found that it pays well to make haste slowly with regard
> to innovations. Electrification for mountain traffic does not carry the same
> appeal that it did two years ago. Oil burning locomotives are solving the
> problem very satisfactorily. Each Mallet compound locomotive, having a
> horsepower in excess of 3,000, hauls as great a load as two of former types,
> burning 10 per cent less fuel and consuming 50 per cent less water.[38]

Reading the occasional news reports, Frank Sprague was uncertain about electrification's future. In a February 1, 1910, letter to William Hood, Sprague expressed his concerns about the press accounts of the new Mallets, and wondered if the railroad's plans had changed.[39] "I never pay any attention to newspaper clippings. I have never known one quoting myself to be correct," replied Hood. "If Mr. Kruttschnitt has decided on any change of plans I am not as yet advised of it."[40]

A July 11, 1910, letter from Sprague to Kruttschnitt asked if his consulting work would continue into the acquisition of locomotives and other equipment for the electrification,[41] and an August response from Kruttschnitt finally let the cat out of the bag:

> we concluded to do nothing except to ask the electric locomotive manufacturers for bids to form an opinion as to just what they could do. Since that time the use of Mallet engines on the mountains, and the uncertainty as to the attitude of the Department of the Interior towards the work necessary to develop our water power, have made us continue to hold the entire matter of electrification of our line over the mountains in suspense, and really indefinitely postpone it.[42]

The decision not to proceed with electrification was, of course, a disappointing one for the electrical suppliers, who had invested substantial work in preparing for the electrification proposals. But it was even more of a personal disappointment for Frank Sprague, for his work as the consulting engineer for what would have been—at that time—one of the world's most demanding railroad electrifications would have enlarged his standing even more as one of the leading electrical engineers and inventors of his time.

In any case, "indefinitely postpone it," turned out to be forever for Southern Pacific's Sierra Nevada electrification. The two articulated Mallet compound locomotives that brought a halt to electrification proved to be the forerunners of an extraordinary steam locomotive fleet. The only problems with the first two locomotives were the poor visibility and the issue of the heavy volume of smoke and gases spilling into the rear locomotive cabs. Since the locomotives were oil fired, it proved feasible to reverse the locomotive boiler with the firebox and cab at the forward end, creating the celebrated "cab forward" design that became standard for almost all Southern Pacific articulated locomotives. Southern Pacific eventually built a total of 244 of these oil-fired locomotives that would carry trains over the Sierra Nevada and other SP mountain grades for almost 50 years until the arrival of diesel-electric locomotives.

Sprague continued in his long-term interest in New York City rapid transit. In 1910 the city was struggling to find a better way to expand its rapid transit system. The nineteenth-century elevated railways served Manhattan, Brooklyn, and portions of the Bronx, while the new Interborough Rapid Transit began operating New York's first subways in 1904 and the Hudson and Manhattan Railroad began operating its subway cars between New Jersey and lower Manhattan in 1908. But the city needed much more rapid transit to accommodate its growing population, extending transit to undeveloped areas of the city, or reducing the density of population in much of lower Manhattan.

There was much disagreement about what to build next, and how. A new Public Service Commission formed in 1907 came up with a plan called the Triborough Subway System. This $150 million system would have three major routes, linking Manhattan, the Bronx, and Brooklyn. Unlike the IRT, which was built with a combination of public and private money, the Triborough's leaders wanted private enterprise to finance, build, and operate the new subways. Bids were finally advertised for privately funded subways in October 1910, but there were no bidders. There were bids for alternately public-private construction, but the PSC did not want to do that. William Gibbs McAdoo, who headed the Hudson & Manhattan, proposed a new extension of his lines to serve Manhattan, the Bronx, and Brooklyn, but the IRT blocked the work by proposing its own extensions.[43]

In the midst of all the uproar about building new subways, Frank Sprague and Oscar T. Crosby, a member of his old Richmond street railway project, joined to propose the development of an Independent City-Built Rapid Transit System. The new plan would include a subway from Times Square south along 7th Avenue and Varick Street, with extensions to the Battery and Liberty Street; a subway across town on 34th Street to Lexington Avenue; a route from 34th Street up Lexington Avenue to 138th Street, with plans for additional connections; and lines linking Liberty Street to points in Brooklyn and the Manhattan and Williamsburg bridges. Proposed financing, which would be much like the earlier subway work for the IRT or the proposed Hudson & Manhattan expansion, was a combination of public funding for rights-of-way and stations, while the private firm would include yard and car housing, and equipment and vehicles. Total costs for the 48-mile system would not exceed $75 million.[44]

Sprague and Crosby submitted their proposal on January 25, 1911, and like so many previous efforts by Frank Sprague to get something done on New

York rapid transit, the proposal was ignored. As it turned out, little was accomplished until 1913, when the new Dual Contracts (joint ventures between the city and the subway builders and operators, the IRT, and the Brooklyn Rapid Transit) were finally organized to build and operate new subways under joint public-private funding.

For Frank Sprague, now 53 years old, this unsuccessful proposal turned out to be his last major railroad electrification project. New electrification projects were largely in the single phase, 25 Hertz AC, or high voltage, overhead wired DC, on which Sprague had not worked in previous projects. Although major North American electrification would continue until the late 1930s, these new electrifications would go to other engineers.

Still working with the wartime Naval Consulting Board in about 1921 when this photograph was taken, Frank Sprague posed with what appears to be some type of naval ordnance equipment.   *Frank Sprague Papers, Manuscripts and Archives Division, The New York Public Library, Astor, Lenox and Tilden Foundations.*

8

THE NAVAL CONSULTING BOARD AND THE GREAT WAR

By the start of the second decade of the twentieth century it became increasingly apparent that war was coming to Europe. Many in the United States realized that despite its isolationism, the United States would eventually have to join the conflict, and that it was neither militarily nor industrially ready to do so, and that furthermore, if steps weren't taken to prepare the country, the United States might be "knocked out" before it could even join in the conflict. This concern became increasingly pressing once war actually broke out in August of 1914. Although it would be more than two years before the United States finally declared war on Germany, in April of 1917, preparations began for that eventuality almost immediately once hostilities broke out in Europe. A significant aspect of these preparations, and in fact the first, was the formation of a civilian Naval Consulting Board.

The board was seen at the time of its formation as a radical departure from the navy's standard mode of operation, but one that was necessary given the navy's need to cope with "the new conditions of warfare"[1] that the European conflict presented as well as the unprecedented threat from submarine warfare. It was the latter, most vividly demonstrated by the torpedoing and sinking of the *Lusitania* on May 7, 1915, that was the immediate impetus for the formation of the board.

Secretary of the Navy Josephus Daniels explained the situation as he saw it in a letter to Thomas A. Edison in early July of 1915.[2] Daniels states in his letter that although the navy regularly received many potentially useful suggestions from the general public, they lacked the infrastructure, personnel, and expertise to systematically develop or evaluate these ideas. What the navy needed, he said, was a body whose express purpose would be to evaluate and implement the suggestions of "our naturally inventive people," as well as generate new ideas of their own. He wanted Edison to head that body. Such a department he said, would "be eventually supported by Congress, with sufficient appropriations made for its proper development" but that for the time being they would "make a start with the means at hand." Daniels felt that public support would be a key element to the success of the board, and for this reason Edison, widely respected and admired as "the one man above all others who can turn dreams into realities," would be the ideal figure to head the board. In his letter, Daniels also stressed the danger of "a new and terrible engine of warfare," the submarine. "The keenest and most inventive minds that we can gather together," along with Edison would be needed to meet this threat. The threat of submarine warfare would be one of the abiding tasks of the board.

Frank Sprague was deeply involved in the process from the very beginning, and in fact suggested the form that the board was to take. Daniels' goal was to staff the board with the "keenest and most inventive minds." Sprague claimed, in a letter to J. C. Parker, that he had suggested the idea to Daniels prior to Daniels' letter to Edison,[3] namely that of contacting the presidents of the 11 leading engineering societies in the United States and asking that each select two candidates from among their membership to join the board.[4] Daniels' original plan had been to let Edison pick four individuals on his own initiative.[5] Edison accepted this change, and work began almost immediately. Over the following few months, the Naval Consulting Board (originally referred to as the Naval Advisory Board) took shape.

Each society provided their candidates through either direct election by the membership of the society, consultation between the society's executives and its members, appointment by the societies' executives, or some other means. Drawn from the worlds of business, industry, and academia, all were prominent in their respective field. The American Aeronautical Society provided M. B. Sellers and H. Maxim; the American Chemical Society, L. H. Baekeland and W. R. Whitney; the American Electrochemical Society, L. Addicks and J. W. Richard; the American Institute of Electrical Engineers, Frank J. Sprague and B. G. Lamme; the American Institute of Mining Engineers, W. L. Saunders and B. B. Thayer; the American Mathematical Society, R. S. Woodward and A. G. Webster; the American Society of Aeronautic Engineers, E. A. Sperry and H. A. Wise Wood; the American Society of Automotive Engineers, H. E. Coffin and A. L. Rikker; the American Society of Civil Engineers, A. M. Hunt and A. Craven; the American Society of Mechanical Engineers, W. L. R. Emmet and S. Miller; and the Inventors' Guild, T. Robbins and P. Cooper Hewitt. In addition to Edison, as president of the board, Secretary Daniels also appointed two members of the War Society of Technical Societies, D. W. Brunton (the chairman) and M. R. Hutchison as members.[6] Over time there was some attrition of board members (Wood resigned within months of the board's creation),[7] but most stayed on throughout the board's existence.

Although Edison had been appointed as president, his role was largely symbolic. The actual work of running the board was largely carried out by Chairman W. L. Saunders and Secretary T. Robbins. Edison's public stature was extremely important in rallying both the members of the board, government officials, and the public to the board's cause, but it also presented problems in implementing the board's recommendations when Edison disagreed with them. This was most evident in the case of the development of the Naval Laboratory (discussed below).

An important dimension of service on the board was that it was voluntary on the part of its members and without remuneration, although after the board was finally legalized and funded by Congress in 1916, expenses of the board members were paid. The voluntary nature of board service, and the implicit patriotism of those who served, was a frequent matter of comment in news coverage and correspondence by and about the board. Another important aspect was that the board was purely civilian, thus it was seen as "disinterested"[8] and therefore free of political pressures for pork barrel spending.[9] Direct involvement of the navy in the board's affairs, particularly the development of the laboratory, would prove to be an extremely divisive issue for the board.

The formation of the board excited intense interest in the press, outside parties, and presumably, the public as well. Newspapers reported on the meetings of the board,[10] and the activities of the board's members, particularly Edison.[11] One such report sparked an angry letter from Sprague to Thomas Robbins, Secretary of the Board, asking him to correct the press's use of the term "Edison Advisory Board."[12] This would continue to be a contentious issue for the board. John R. Dunlap, editor of *Engineering Magazine* and an early promoter of management science, ran a lengthy feature on the board in the November issue of his magazine,[13] and wrote vehement letters to Secretary Daniels,[14] Frank Sprague,[15] and presumably other members of the board, urging them to get to work immediately. Overall, there appears to have been a public spirit of high optimism for the board's prospects.

The board held its first organizational meetings on October 6–7, 1915 (just a little over a year after the start of the war, and five months exactly after the sinking of the *Lusitania*) with almost all members present.[16] They adopted rules of governance and were commended for their efforts by President Wilson. At a meeting the following month (November 4) they divided their work between 16 committees (later increased to a total of 21 committees, listed below).[17] Sprague distinguished himself by serving on five committees, including Electricity (which he chaired), Submarines, Ordnance and Explosives, Ship Construction (which he chaired), and Special Problems (although not appointed to the latter until 1917).

When it was initially created in October of 1915, the board had no legal mandate, source of funding, or even formal articulation with the various naval bureaus—it was simply attached to the office of the Secretary of the Navy. This led to a number of operational challenges, including the question of how the board's expenses would be met. Members received regular messages from board secretary Thomas Robbins regarding their share of operating expenses.[18] There also is a great deal of correspondence discussing organizational issues and a variety of problems that board members faced, from being ignored by the navy,[19] lacking proper direction from the navy,[20] to the poor quality of the stationery.[21] Many, though not all, of these problems were resolved when the board was finally legalized by an act of Congress (H.R. 15947) on May 26, 1916. The members were formally sworn into office on September 19, 1916,[22] and allocations were made for board expenses and the establishment of a naval laboratory. The site and constitution of the laboratory would be a matter of great importance to Sprague and the other board members, and yet another issue that proved to be extremely divisive.

Although the ultimate success and effectiveness of the board is debatable, as we shall see presently, the board certainly had problems and conflicts of personality that attenuated its effectiveness, and these were apparent almost from the beginning. As early as December of 1915, a mere two months after the formal inauguration of the board, Sprague was at pains to point out to his colleague Lawrence Baekeland, who had in a comment to the press incorrectly credited Thomas Edison with the conception of the constitution of the board, that he (Sprague) and not Edison had proposed to Secretary Daniels the final form that the Consulting Board was to take.[23] Baekeland promptly replied, asking Sprague's pardon,[24] but these and other more substantial disputes (including those mentioned above) continued to dog the board well past the end of the war and, in fact, past its ultimate dissolution.

The problems facing the board in late 1915 were manifold. They ranged from issues such as improving internal combustion engines, reducing friction on the skin of ships, and determining oil flow rates through various sizes of pipes, to electric detonation of mines, reducing erosion in gun barrels (naval guns), and of course the submarine menace.[25] The board's initial list of problems was two pages long.[26] Beyond the number of individual technical issues the board had to deal with, the very nature of the problems they had to tackle was quite broad as well, dealing with both issues that required the improvement of existing technologies (e.g., mitigating or preventing erosion in gun barrels) and issues that required the development of entirely new technologies (e.g., the detection of submarines). They chose to approach their task by focusing on specific problem areas, and eventually established 21 standing committees to address them. These were:

Aeronautics, Including Aero Motors: E. A. Sperry (chair), B. J. Arnold (Arnold had worked with Sprague previously on the New York Central Electrification), H. E. Coffin, P. C. Hewitt, A. L. Rikker, M. B. Sellers, and A. G. Webster.

Aids to Navigation: E. A. Sperry (chair), A. Craven, A. M. Hunt, and R. S. Woodward.

Chemistry and Physics: W. R. Whitney (chair), L. Addicks, L. H. Baekeland, J. W. Richards, M. B. Sellers, A. G. Webster, and R. S. Woodward. The committee was later split between *Chemistry* (chaired by W. R. Whitney) and *Physics* (chaired by A. G. Webster), both with largely the same membership as the original committee.

Electricity: F. J. Sprague (chair), L. Addicks, W. L. R. Emmet, P. C. Hewitt, B. G. Lamme, and A. G. Webster.

Fuel and Fuel Handling: S. Miller (chair), L. Addicks, L. H. Baekeland, A. M. Hunt, M. R. Hutchinson, H. Maxim, T. Robins, B. B. Thayer, A. G. Webster, and W. R. Whitney.

Food and Sanitation: L. H. Baekeland (chair), H. Maxim, B. B. Thayer, and W. R. Whitney.

Internal Combustion Motors: A. L. Riker (chair), H. E. Coffin, M. B. Sellers, and E. A. Sperry.

Life Saving Appliances: S. Miller (chair), H. Maxim, and T. Robins.

Metallurgy: J. W. Richards (chair), L. Addicks, B. G. Lamme, B. B. Thayer, and H. Maxim.

Mines and Torpedoes: E. A. Sperry (chair), L. H. Baekeland, M. R. Hutchinson, and H. Maxim.

Optical Glass: L. H. Baekeland (chair), J. W. Richards,
   A. G. Webster, and W. R. Whitney.
Ordnance and Explosives: H. Maxim (chair), L. H. Baekeland,
   A. M. Hunt, M. R. Hutchinson, F. J. Sprague, A. G.
   Webster, W. R. Whitney, and R. S. Woodward.
Production, Organization, Manufacture and Standardization:
   H. E. Coffin (chair), L. Addicks, W. L. R. Emmet, B. G.
   Lamme, T. Robins, W. L. Sanders, and B. B. Thayer.
Public Works, Yards and Docks: B. B. Thayer (chair), L. Addicks,
   A. Craven, A. M. Hunt, S. Miller, and J. W. Richards.
Ship Construction: F. J. Sprague (chair), A. M. Hunt, M. R.
   Hutchinson, S. Miller, and J. W. Richards.
Special Problems: B. G. Lamme (chair), L. Addicks, A. M.
   Hunt, M. R. Hutchinson, M. B. Sellers, E. A. Sperry, F. J.
   Sprague, A. G. Webster, and W. R. Whitney.
Steam Engineering and Ship Propulsion: A. M. Hunt (chair), W. L. R.
   Emmet, B. G. Lamme, J. W. Richards, and M. B. Sellers.
Submarines: W. L. R. Emmet (chair), A. M. Hunt, M. R.
   Hutchinson, W. L. Saunders, and F. J. Sprague.
Transportation: B. J. Arnold (chair), H. E. Coffin, A. Craven, S. Miller,
   A. L. Riker, T. Robins, W. L. Saunders, and B. B. Thayer.
Wireless and Communications: P. C. Hewitt (chair),
   A. G. Webster, and W. R. Whitney.

In addition to the standing committees, there were several ad hoc committees, the most important of which was the Sites Committee, which was charged with planning the laboratory.

However, before any of these problems could be addressed, an even larger, more fundamental issue had to be tackled: America's industrial preparedness. Lloyd N. Scott, the official historian of the Naval Advisory Board, contrasts German industrial preparedness at the start of the war—they were able to shift from civilian to military production in a matter of hours—to that of the United States, where even the capacity of production was unknown.[27] Before the board had even been legalized and given financial support from Congress, they began an Industrial Preparedness Campaign. Their first step was to conduct a survey of industrial capacity throughout the country, which would then guide them in planning America's wartime industrial efforts. They accomplished this through a system similar to the one through which the board members themselves were selected. They canvassed the members of five of the engineering societies that had provided the board's members, by state, to form state committees which would carry out an industrial preparedness survey for their state. This operation cost tens of thousands of dollars overall, with the costs paid by the participating members.[28] The effort was headed, on the part of the board, by Howard Coffin (as chair of the committee for Production, Organization, Manufacture and Standardization), who a few years previously had been a leading force behind the effort to introduce standardization to the automotive industry. The campaign also benefited from the eager participation of other figures such as President Wilson and Thomas Edison, who helped to popularize the campaign.[29] The Industrial Preparedness Campaign accomplished its goals in a remarkably short period of time and ultimately, their efforts led to the creation of the Council of National Defense.[30]

Once the board had helped to put the American industry on a war footing, the board members turned their attention to the other tasks that they had appointed themselves, although the priority they gave to various tasks and the intensity or urgency with which they approached them varied as the board went from the pre-war period, through the war, and into the post-war period. The work fell into two broad domains: first, as Secretary Daniels had suggested in his letter to Edison, evaluating and, if merited, implementing ideas sent in by the public (discussed in the next section); and second, a more self-appointed task, that of identifying specific problems or needs of the navy and addressing them directly.

The board addressed the latter issue head-on in January of 1916 when they sent Frank Sprague on a cruise on the USS *New York* to make a first-hand evaluation of the fleet. As a former naval officer, he was the ideal member for such a mission, although over the course of the board's existence, most members would at one time or another make similar, if smaller-scale efforts. Sprague went on his cruise under intense press coverage; "Sprague Sails With Fleet"[31] and "Civilian Expert Guest on U.S. Dreadnaught"[32] were typical headlines. After spending a month aboard the USS *New York*, Sprague's findings were somewhat equivocal. While not wishing to be too critical, he was clear to point out that there were many improvements to be made. His report was subject to extremes of interpretation: "U.S. Navy Inferior, Says F.J. Sprague,"[33] "No One Need Feel Ashamed of the Navy, Says Sprague,"[34] "U.S. Has Nucleus of World's Best Navy, Says Expert,"[35] and "Frank J. Sprague Has High Praise for American Navy"[36] capture the range of opinion.

Although attention from the newspapers continued to be intense through the war years, much of the board's work went largely unnoticed by the public. In fact, much of the work was quite unglamorous: dealing with ideas submitted by the public, often to the wrong committee or department; forwarding misdi-

Frank Sprague generated wide publicity early in 1916 when he was sent on a month-long trip on board the dreadnought USS *New York* to evaluate the fleet, bringing back a variety of recommendations for the navy. Profiting from his experience, for example, Sprague came up with an idea for a new system of "directorscope," which would enable the simultaneous operation of all main guns under automatic control from a periscope in the conning tower or elsewhere, allowing the *New York* to fire a salvo of all ten of its main guns. *Naval Historical Center (Neg. 19-N-13046).*

rected submissions from one committee to another; evaluating and reporting on public submissions; and on occasion, performing actual experimental work on a particular problem. Lloyd N. Scott singles out the board's work on fuel oil, the special problems committee (largely submarine oriented), and ship protection for special notice.[37]

The fuel oil issue was of considerable import. At the time, the navy had ships that burned both coal and fuel oil in their boilers. Coal was inexpensive and domestically abundant whereas the supply of oil was precarious, and the navy was in danger of losing its strategic oil reserves. On the other hand, coal, especially bituminous coal, produced much more smoke than oil did when it burned, and therefore could reveal a vessel's or convoy's position.[38] The board was asked to consider the question from the point of view of logistics and infrastructure (i.e., the production, transportation, and storage of various fuels) as well as the mechanics of their utilization (i.e., design requirements, performance parameters, and other aspects of the various machinery).[39] In addition to fuel oil, the board also investigated the use of pulverized coal.[40]

Because different aspects of the submarine problem were handled by different committees (e.g., submarines, mines and torpedoes, etc.), the board established yet another committee, special problems, to coordinate all submarine-related issues. The special problems committee's members included Lawrence Addicks, M. R. Hutchinson, B. G. Lamme, M. B. Sellers, E. A. Sperry, A. G. Webster, W. R. Whitney, and A. M. Hunt. They attempted to tackle the problem from several angles, including detecting submarines acoustically, electromagnetically, visually, and by underwater searchlight, attacking submarines from the air, protection from torpedoes, and so on. They also organized a conference on submarines in early March of 1917 in which they identified the three major issues in coping with the submarine menace: detecting submarines, attacking them, and defending ships against torpedoes. The committee in particular dealt with the issue of detection. Working in conjunction with several other committees and naval bureaus they were successful in developing, testing, and deploying several different acoustic submarine detection systems. They also investigated several magnetic and electronic systems that were less successful.[41]

The board also worked with other entities, including the Ship Protection Committee and the United States Shipping Board, on the problem of protecting merchant vessels from submarine attack. It was widely recognized that Germany would win the war if the Allies could not be re-supplied by trans-Atlantic shipping. A wide range of strategies were investigated, from nets and shields to protect vessels from torpedoes, to various techniques to disrupt or destroy torpedoes at a distance; passive defenses such as smoke screens, camouflage, and profile disrupting paint-schemes; and convoying strategies. Passive systems, particularly convoys, ultimately proved to be the most effective means of protecting shipping.[42]

Aside from these specific accomplishments, the board also played the role that Secretary Daniels had envisioned for them: serving as a clearinghouse for inventions submitted by the public.

In reviewing the board's correspondence, one is struck not only by the volume of public submissions, but also by the expectations that the public had that their suggestions would be given prompt, if not immediate, attention and be acted on in due course. There are numerous exchanges, some covering periods of mere days, between Sprague and various public-spirited inventors in which he explains, in response to their sometimes highly irate queries, why their invention has not been adopted, that the review process takes time, and that the board was receiving hundreds of submissions per day (the all-time high was 600 in a single day and a grand total of about 110,000).[43] It seems that many expected a response by return mail and immediate action on their invention.

The majority of the submissions to the board were variations on a theme (e.g., anti-torpedo devices, unsinkable ships, new designs for submarines, etc.), and most, as might be expected, were naïve and plainly impractical, although well intentioned. Inventors saw the problems clearly enough, but lacked the basic knowledge to fully understand or practically address them. There were, for instance, many suggestions that ships be protected from torpedoes by nets or shields suspended from booms mounted on the ship's sides, not realizing that the drag this produced would slow a ship to a dead crawl—nets were used, however, to protect stationary ships—or that glass-bottomed boats equipped with underwater searchlights to detect submarines would be ineffectual because underwater visibility is extremely limited due to the water's opacity. In fact, most were not inventions at all, but simply concepts or ideas that had not been tested or developed, and which often were poorly illustrated and described. Some submissions were from certifiable crackpots who proposed a variety of harebrained schemes whose basic premises violated the laws of nature or were simply unimplementable—for instance cork pads mounted on springs attached to a ship's hull to cause the torpedoes to bounce back. A few submissions were self-serving, self-promoting, or merely venal, but these seem to have been rare. Others were highly public-spirited offers to put an individual's or organization's capabilities at the service of the government, such as an offer to make the students and facilities of a technical school available for the war effort.[44] Overall, the range of submissions presents a picture of a concerned and patriotic public who wished to do what they could to help their country in time of need.

Submissions went through a multiple-stage vetting process in which ideas that seemed to have merit percolated upward to the members of the board's

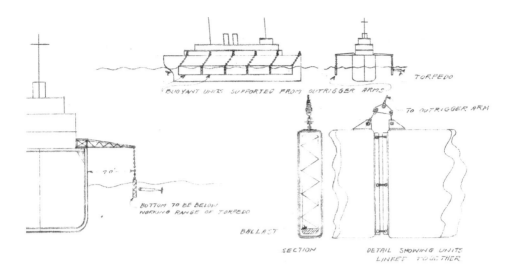

various committees that dealt with the specific area of the submission. While many of the least useful ideas were eliminated through this process, the volume of material that the members of the board had to deal with still was considerable. Each submission was patiently answered, at least during the board's early days, with an individual letter explaining why the idea was not being adopted, but thanking the inventor for their help. This often entailed subsequent letters dealing with the inventor's protestations of the rejection of their idea, requests for collaboration, or even personal visits by the inventor. In the later phases of the board's career a form letter was adopted for rejections. This responsibility was taken quite seriously by the members of the board. There are several instances in Sprague's correspondence in which one board member or another is enquiring about the fate of a particular submission, and occasional admonishments from Thomas Robbins, the board's secretary, to promptly deal with public submissions, forward them to the appropriate committee if misdirected, and return original materials when so requested.

In spite of Secretary Daniels' optimism, however, very few of the submissions from "our naturally inventive people" proved to be practical or original, and most importantly, few could be implemented within the span of the war. Lloyd N. Scott estimates that of the 110,000 submissions received, only 110 had merit, and of these, only one was implemented during the war, the Ruggles Orientator, a flight simulator used for training pilots.[45] Scott himself is quite dismissive of the very notion of soliciting inventions from the general public, which he, perhaps rightly, regards as simply lacking the necessary knowledge and training to deal competently with the complicated issues of modern naval warfare.

SPRAGUE'S CONTRIBUTIONS TO THE BOARD

It is clear from Sprague's correspondence that he took his work with the board extremely seriously, and that he was repeatedly, if not constantly, frustrated by what he saw as limitations on the board's scope and impact. At various instances he threatened to resign, and at one point actually did so, although his letter of resignation was discreetly overlooked.[46] Perhaps his greatest frustration was that the board never lived up to what he thought its potential to be.

Sprague's work on the board through the end of the war was a combination of board meetings, committee meetings, reviewing inventions submitted by members of the public, implementing a variety of plans, and on occasion, experimental work. Of these responsibilities, he was least interested in reviewing the public's "inventions." On a number of occasions he wrote to his colleagues of his frustrations with having "to act as a clerk to comment on other people's inventions" and having no time for any of his own,[47] although his letters to the submitters were always polite and patient, even when he clearly thought they were impractical.

His service on several committees, Electricity, Ordnance, Ship Construction, Special Problems, and Submarines, gave him wide-ranging responsibilities and diverse problems to deal with beyond his review of public submissions. In addition, he had his personal business interests to attend to. These included serving as a consulting engineer of the Sprague Electrical works of the General Electric Co., President of the Sprague Safety Control and Signal Corp., and President of the Sprague Development Co.[48] Also, throughout his entire time with the board, he was engaged in a lengthy and very demanding legal fight over his automatic train control patent application.

Sprague's work on the board got off to a very public start with his voyage with the winter cruise of the Atlantic fleet onboard the USS *New York* for four weeks in January and February of 1916. Unfortunately, this in some ways might be seen as the highpoint of his tenure on the board. A press release by Secretary Robbins stated: "If Mr. Sprague is able to make as great scientific contributions to the Navy as he has already given to the traveling public, it would be impossible to overestimate the amount by which the country's defenses would be strengthened." The entire voyage was subject to intense press coverage and public interest, and many newspapers featured dramatic photos of artillery salvos during target practice. At the end of the voyage, Sprague submitted a report to Josephus Daniels (secretary of the navy) detailing his findings.

His participation in the cruise had been in his capacity as chair of both the committees on Electricity and Ship Building. While he found that he had insufficient information to make any recommendations on shipbuilding, he had a number of recommendations for the use of electricity in communication, lighting, signaling, industrial motor application, gun fire, and electric steering.[49]

Following the voyage on USS *New York,* Sprague's work settled into a routine of committee meetings, review of proposals from the public, and on occasion experimental work (in addition to his private business, which went on apace during this period). Judging from his correspondence, his work covered a vast scope: issues included defense of merchant shipping from submarines (both active defense and tactics); means of combating submarines (weapons, tactics, detection, submarine hunting vessels); a wide range of ordnance issues (gun erosion—the degradation of canon barrels due to use, depth charges, torpedoes, mines, and the production of explosives—including efforts to produce nitrogen from air); propeller design; improvements to the Naval Academy; survival gear for lifeboats; helicopters; advertising (propaganda); and the use of pulverized coal as fuel.

While it is impossible to outline the details of all of these projects, a few stand out as particularly worthy of note. There are several proposals for an electric propellant gun, or simply, an electric gun. They outline what essentially is a rail gun, a weapon that uses electromagnetic energy to launch a metal projectile with an extremely high velocity and massive destructive impact.[50] Rail guns are familiar to fans of science fiction and video games, and, surprisingly, have been explored since the mid-nineteenth century as actual weapons.[51] Although technical problems have prevented the realization of the theory until recently—the U.S. Navy completed successful testing of a rail gun in 2007 that when fully developed will be capable of launching a projectile over 200 miles—the French were actively experimenting with the concept during the First World War.[52] The board found the proposals to be impractical, but one of the proposers, F. B. MacLaren, was eventually awarded a patent for his invention (no. 1,384,769).

Another interesting project in which Sprague was involved was a long-range, low-speed torpedo. This was essentially a self-propelled mine that could be released at a distance from its target and then make its way in the desired direction. The concept was that large numbers of these mines could be released off the German coast and block access by the German navy to the Atlantic. What is most interesting about this plan is Sprague's insistence that the mines be equipped with a timing device that would disable the mines after a set period of time so that they not become a hazard to general navigation.[53] In view of the terrible toll that mines of all types take on civilian populations today, such a precaution was indeed merited.

Again, judging from his correspondence, it seems that the majority of Sprague's work on the board was focused on the general subjects of submarine warfare, in all its various manifestations, ordnance, and shipbuilding. Given Sprague's various committee assignments, the committee on Special Problems in particular (of which he was made a member on May 23, 1917),[54] this isn't surprising. The board attacked the problem of submarines from a variety of angles, from defending merchant shipping from attack, to attacking submarines and developing the American navy's capacity to launch submarine warfare in its own right.

Sprague was acutely aware of the danger that submarine warfare presented to the viability of the war effort. In a letter to President Wilson of July 20, 1917, he stated: "The submarine menace is the supreme peril of the hour. Unless it can be controlled within the year there is no end of the conflict now in sight and the ultimate issue will be in doubt." He went on to cite the statistics on sinkings for the year: 700,000 tons a month, 1,000 tons an hour.[55]

He reviewed a multitude of proposals for dealing with the submarine menace, the vast majority of which could best be described as ill informed on the technical requirements and basic physics of the problem. These proposals included the use of protective nets or booms, grappling hooks, mirrors, smoke bombs, rapid pursuit vessels, detection systems, mines, a "collision mat" that could be used to prevent a torpedoed ship from sinking, and so on.[56]

Sprague's own efforts were multifaceted, focusing on both defensive and offensive measures. He states in his "Proposal for the Defeat of Submarines" that this "menace will not be overcome by any single, mysterious invention, and the frequent newspaper references to such, 'infallible' and otherwise, and quotations cited as emanating from distinguished scientists, are both misleading and injurious." He goes on to state that the "submarine peril cannot be met merely by defensive measures or by feeding it with a bridge of wooden boats or a fleet of steel ones," and that it is "a fundamental rule of war that conclusive victory depends upon and is won only by attack and not by defense."[57]

His plan relies on several interrelated elements:

1. Contrary to the current practice in which "merchant men play hide-and-seek with submarines" on the open sea, definite sea lanes should be established.
2. The sea lanes would be guarded by a new fleet of specially constructed "submarine destroyers," vessels that should be small (about the size of a North Sea trawler), fast, maneuverable, oil-burning, steam driven, and shallow draft (rendering them less vulnerable to torpedo attack).
3. The vessels would be armed with depth bombs and dwarf torpedoes (both novel inventions that Sprague was working on in one capacity or another).
4. They would be supplied by a fleet of mother ships, "made practically immune to submarine attack," that could re-provision the submarine destroyers.

While Sprague was at pains to point out that the larger, standard destroyers were "splendidly effective," he found that they were too slow for the vigorous patrols he envisioned and too costly to produce in the necessary number his proposal calls for. The plan also contains specifications for two possible boat designs and the dwarf torpedo.

Sprague fleshed out his proposal with a detailed analysis of data on ship sinkings from several sources, including the German government. While his sources disagree in specifics, he points out that they all present an appalling picture of loss of life and material. Moreover, even with increased American production of merchant shipping, he points out, the rate of loss would be simply unsustainable in the long run. He found that even recent advances in combating the submarine menace, such as camouflage, smoke screens, and improvements in weaponry, while reducing losses, were at best a "palliative," since they did not "go to the root of the evil."

Sprague submitted his proposal to both Secretary of the Navy Daniels and President Wilson in July of 1917. The navy received it, at best, with skepticism. In spite of repeated letters to Secretary Daniels with efforts to correct misunderstandings of the proposal and data to show the practicality of its various elements, his plan was rejected. The navy had already committed to the construction of a new fleet of 150 larger destroyers.[58] The rejection of his proposal was a matter of some unhappiness for Sprague.[59]

Sprague's second major area of work was ordnance. As a member of the Ordnance Committee he reviewed numerous public proposals and worked on committee business regarding the fixation of nitrogen from the atmosphere (for use in explosives—there was a fairly high level of concern that explosives manufacture would be imperiled if access to Chilean nitrate deposits should be cut off),[60] gun erosion, and a variety of mines, bombs, and torpedoes. Sprague's work in ordnance also allowed him a few opportunities for what he seems to have enjoyed most: developing and testing his own ideas. The ordnance work also allowed him to work closely with his son Desmond, who by then also worked with his father in the Sprague Safety Control & Signal Corp.

In a letter to Lloyd N. Scott, Sprague outlines his work in ordnance.[61] He describes his work on two anti-submarine net systems to which bombs were attached, aerial bombs with leaders (which would detonate the bomb above ground, causing greater destructive impact), and a variety of fuses for depth charges. He points out that a number of these were very similar to already patented ideas, but he underscores that all of these preexisting patents depended on methods that he had proposed many years earlier. He also describes ongoing work on a variety of delay-action fuses, the idea for which had been proposed by his son Desmond.

Sprague and Desmond worked on their depth charge and other fuses between 1917 and 1920. Their most intensive period of work seems to have been in 1917 and 1918, when they conducted a series of tests in Indian Head, Maryland, and Newport, Rhode Island.[62] Sprague financed much of his and Desmond's work out of his own pocket, some of which was eventually reimbursed by the board.[63] Sprague also received funding from the Submarine Defense Association for some of his tests.[64] This latter source of funding seems to have caused some small measure of trouble for Sprague: at one point he found it necessary to point out to Rear-Admiral Ralph Earle of the Naval Bureau of Ordnance[65] that although he received private funding for some of his research, he had not "allowed the question of commercialism to intrude itself."

Ultimately, the navy rejected his proposals for depth charges and fuses, in spite of a fairly laudatory report from the Special Board on Naval Ordnance in which both their safety (multiple features including a failsafe that would prevent depth charges from becoming a hazard to navigation if they failed to explode as planned) and stability are cited.[66] There seem to have been some questions as to the accuracy of the report, as well as problems with patents. As Sprague pointed out in his letter to Scott, the patents in question had all been based on his earlier work, but aside from making the point, he declined to engage in any "personal controversy" over the matter.[67] Despite their rejection by the navy, Sprague and Desmond continued their work on depth charges and fuses, and Desmond (with Phillip Whaley Allison) eventually was granted three ordnance-related patents (nos. 1,343,415—a fuse; 1,382,750—a projectile; and 1,512,249—a firing mechanism).

With the exception of Sprague's proposal for the construction of a fleet of new submarine destroyers discussed above, his work on shipbuilding was mainly confined to the review of proposals and routine committee work. He reviewed an enormous number of proposals for everything from wooden boats, concrete boats, unsinkable boats, mosquito boats (seemingly forerunners to the patrol torpedo or PT boat), and shallow draft coastal defense boats, to a variety of submarines, including one-man, two-man, and full-scale models. He also reviewed problems particular to submarine design, such as the absorption of hydrogen released by submarine batteries—an extreme safety hazard—and the use of bottled oxygen to allow submarine's diesel engines to run while submerged. He dealt with many topics relating to marine technology, including turbines, gears, propellers, electric motors, watertight doors (for which he held several patents), the use of pulverized coal for fuel, and the like.

Sprague's final major contribution to the board was his work on the Naval Laboratory. The laboratory was without doubt Sprague's greatest source of aggravation during his period of service on the board. Over time it seems to have come to embody everything that, in his opinion, had gone wrong with the board.

THE NAVAL LABORATORY

At the time that America was preparing for the war and the board was being established, the navy had several facilities for experimental work and testing, although none of these could properly be called a research laboratory. These facilities included, under the Naval Bureau of Ordnance, the Torpedo Station in Newport, Rhode Island (at which Sprague and Desmond conducted a number of tests in 1917 and 1918); the Naval Gun Factory in Washington, D.C.; the Naval Proving Ground in Indian Head, Maryland (where Sprague and Desmond also conducted tests); a chemical laboratory in the Philippines (which tested gunpowder for stability); and the USS *Montana* (used to test torpedoes and torpedo tubes).[68] The Bureau of Construction and Repair maintained an experimental wind tunnel and a model basin at the Washington Navy yard and several laboratories for mechanical testing at the New York yard.[69] The Bureau of Steam Engineering maintained an Engineering Experiment Station in Annapolis, Maryland, for making efficiency tests on various equipment as well as fuel and metal, an Oil Fuel Testing Plant at the Philadelphia yard, an Electrical Testing Laboratory at the New York yard, several Radio Laboratories, an Aeronautic Engine facility at the Washington yard, and several Chemical Laboratories at various yards.[70]

Collectively these facilities constitute an impressive range of research capabilities for a navy at the start of the twentieth century. Individually, however, they are either limited in focus, like the torpedo station at Newport, or limited in capability, like the oil fuel testing lab at Philadelphia. In addition, they were distributed over three separate naval bureaus and already engaged in the work for which they had been established. Clearly, none of these facilities were capable of the broad range of issues that the Naval Advisory Board was charged to investigate, and so the Naval Laboratory was one of the board's earliest recommendations. L. H. Baekeland was an early advocate of a Naval Laboratory, writing an enthusiastic call for its development in December of 1915.[71]

Although the Naval Laboratory came to dominate Sprague's work on the board (as well as that of other members), especially in the post-war years, he does not seem to have given it a great deal of thought at the start of his tenure on the board. In 1915, before the board had been legalized, Sprague points out in a series of letters with Secretary Robbins that concerns about the laboratory's location are premature, as it hasn't been approved nor have any allocations for the laboratory been made. Later the same year, he declined an opportunity to chair the Laboratory Subcommittee, as he planned to be away on his maneuvers with the USS *New York*.[72] He clearly imagined that establishing the laboratory was going to be a straightforward proposition that would quickly be dispensed with, and had no inkling that this was to be the board's most divisive and frustrating undertaking.

Sprague began his advocacy of the laboratory shortly after his voyage aboard the USS *New York*. He points out in a letter to Secretary Robbins that, given the navy's limited facilities, much of the experimental and testing work necessary for the board's investigations would have to be carried out privately, but that should the navy have a proper research facility, the work could be greatly furthered.[73] Until the board was legalized and allocations were made for its operation, however, the laboratory remained purely hypothetical. This changed a few months later with the legalization of the board and allocation of funds for the laboratory in May 1916, and the board immediately set to work on their proposal for the laboratory. They also, almost immediately, ran into a host of unforeseen obstacles, some of which would dog them for the next six years—well into the post-war period.

One of the immediate obstacles to their proposal was purely political. Although many had hoped that because the board's members were private citizens and volunteers, their work would be free of pork barrel politics, and on an individual basis, this seems largely to have been the case. However, by its nature, their work was in the political arena, and so could not altogether avoid the pork barrel. Creating a laboratory meant situating it somewhere, and so hosting the laboratory became a coveted prize.

Even the initial bill (HR Bill 9702) for making allocations for the laboratory was imperiled by pork barrel politics. One of the bill's sponsors, Representative Walter R. Stiness of Rhode Island, originally included language that favored (though did not demand) Narragansett Bay as the location for the laboratory. He was persuaded to withdraw the locational preference after consulting with L. H. Baekeland and Sprague who wished to avoid the "pork barrel flavor" of such language.[74]

Throughout the fall and winter of 1916, Sprague (who by this time had joined the subcommittee on sites) received numerous entreaties from various localities, their boosters, or their representatives in Congress suggesting theirs as the ideal location for the laboratory. In one instance, the Maryland Electric Railway offered to extend their Annapolis line to the proposed site of the laboratory as an inducement.[75] He also received a series of letters from W. F. M. Goss, Dean of the College of Engineering at the University of Illinois at Urbana Champaign, proposing a "segregated" laboratory, in which facilities could be located at the University of Illinois and elsewhere (probably due to Urbana Champaign's notable lack of port facilities).[76] The board disregarded these requests and meticulously carried out their search for a laboratory site that met their criteria.

A second, significant obstacle was financial. The bill authorizing the laboratory allowed 1.5 million dollars, but only allocated 1 million. The board had all along envisioned an initial laboratory budget in the neighborhood of 5 million dollars.[77] Lack of funds was a persistent problem not just for the laboratory, but for all of the board's endeavors.

A third, perhaps surprising source of resistance to the laboratory was the navy itself. Although not too apparent at first, the navy never really was enthusiastic about the laboratory. This is not as surprising as it might seem: at this time the navy was preparing for, and would shortly be engaged in, the First World War. They had numerous wartime expenditures and a limited budget, so seeing 1 million dollars go to a research laboratory that would be largely under the control of civilians cannot have been appealing. Their opposition to the laboratory became apparent when the board made its report on potential sites in December of 1916.

The final and most confounding obstacle to the establishment of the Naval Laboratory was a member of the board itself: its president, Thomas Edison. Edison very early in the site selection process fixed on a single acceptable location for the laboratory, Sandy Hook, New Jersey, and simply refused to budge. Later, he also added his opposition of any direct involvement in the laboratory by the navy, insisting that it be under sole civilian control. In spite of being the only member of the board in opposition to the majority opinion, he maintained his stance through the close of the board.

Edison's intransigence presented several problems. First, his steadfast resistance to the report of the Committee on Sites in a sense tied the hands of Navy Secretary Daniels. He could scarcely publicly disregard the opinion of his iconic and personally appointed Board President, regardless that he was a minority of one. Secondly, Edison's refusal to back the board's plan provided a pretext for naval officers (who were less than enthusiastic about a civilian-led naval laboratory) to put off the entire project until after the war, which is precisely what happened.

Edison's willingness to go his own way was apparent from the start. In the late summer and early fall of 1916 the Committee on Sites began making inspection visits to various potential sites. Edison did not accompany them, but made his own visits. He also presented his own report on four prospective sites in October of 1916.[78] In his report, he found only one truly suitable candidate, Sandy Hook, to which he steadfastly held for the duration of the laboratory debacle.

Edison's contrarian stance was not lost on his colleagues. The other members of the Committee on Sites quickly realized that they had an implacable foe in Edison, and one who was potentially very destructive to their mission of establishing a naval laboratory. In early December 1916, there were a series of exchanges between various members of the Sites Committee urging unanimity in their selection of the site (Annapolis) and how they were to either get Edison to join them or to quash his proposal for Sandy Hook. These letters occasionally verge on the conspiratorial and included substantial coaching for how Sprague might preempt Edison's views.[79]

Sprague had been tasked to write the official report of the Committee on Sites, and into mid-December of 1916 was hard at work on the task. He circulated drafts of the report to the other members of the committee, with the notable exception of Edison. Apparently, tensions had risen to the point that

Sprague felt that he could gain nothing from Edison's input. He did however receive ample feedback from the other members of the committee, which he clearly incorporated into the final version. Edison finally requested a copy of the final report through Secretary Robbins,[80] and Sprague complied, noting in the accompanying letter to Edison that the report contained arguments both for and against Sandy Hook, and expressing the hope that Edison's support could make the Committee's report unanimous.[81] Edison submitted a minority report, which Sprague felt was "hardly according to Hoyle" given that his views had been incorporated into the Committee's report.[82]

Sprague had been working on the report for several months with his fellow Committee members (with the apparent exception of Edison), and over this period they managed to forge a unanimous opinion about the nature and composition of the proposed laboratory as well as its location (again with the exception of Edison). In the course of developing the report, they formally reviewed 45 proposed locations from New Hampshire to Louisiana[83] and met with naval officers to gather their input.[84]

In his report, Sprague cites the enabling legislation and the shortfall in funds between what the board had requested and those actually allocated. He points out that, as a result, financial constraints had to be a factor in the Committee's recommendations. He proposes that the laboratory be under the direction of a single, high-ranking naval officer who could act as an intermediary between the various Naval Bureau heads and the Naval Consulting Board, avoiding the dangers of creating a "many-headed, inefficient organization." He also recommends that the laboratory be set up on the model of an experimental laboratory with more extensive equipment than currently was available in the existing naval laboratory facilities.[85]

After presenting a general overview of the laboratory, Sprague tackled the issue of the actual site. The general parameters were the physical suitability of the site and minimizing cost and time necessary to establish the laboratory. Sprague laid out the differences between Edison and the Committee, presenting Edison's views first and those of the Committee second. The differences were both in their conceptualization of the laboratory (Edison preferred more of a fancy machine shop while the rest of the Committee a research laboratory) and the location: the Committee found Sandy Hook inferior on a number of grounds, such as its inconvenient location, its being subject to naval bombardment, being owned by the army, and so forth. Sprague then outlined the favorable qualities of the Annapolis location (the final, unanimous choice of the remaining board members). These included everything from the availability of the land (already owned by the navy), the depth of the channel, its proximity to Washington, D.C., availability of labor and materials, and even the quality of local housing. He enumerated 18 favorable points in all. He also pointed out that "Annapolis is the seat of the Naval Academy, and the prime source of the professional education of the officers who are especially concerned," and cited the advantages that this arrangement would confer (what today is commonly called synergy).[86]

The Committee's report was submitted to Navy Secretary Daniels on December 14, 1916, by Board Secretary Robbins. In due course, the proposal was rejected by the navy. Although the navy's preference that the laboratory be located near Washington, D.C., played a significant role in their decision, the fact that Edison had so vehemently opposed Annapolis also was a significant factor. While various members complained from time to time about the lack of a laboratory (particularly Sprague), the board did not actively revisit the subject for more than a year. Sprague didn't cease in his personal efforts, however. In a letter to Senator John W. Weeks regarding an interview that the senator had arranged between Sprague and one of Weeks' constituents, Sprague mentioned that the constituent, one Mr. Cummings, was the ideal sort of person to be working in the Naval Laboratory, but that, unfortunately, it still did not exist despite a congressional allocation.[87]

The board made a second report on the laboratory in June of 1918, this time proposing the Bellevue Magazine, located on the Potomac River and near the Washington Navy Yard, as the location.[88] In advance of the second report, Sprague wrote to Edison, trying to persuade him that, in light of the war, the laboratory was now a vital necessity.[89] However, Edison again vehemently opposed the proposed location as well as anything but the most superficial involvement by the navy. Although Bellevue was the navy's preferred location (and in fact, where the laboratory eventually was built), Edison's continued opposition persuaded them to again reject the plan. The fact that the navy was busy fighting the war probably was a significant contributing factor.

With the end of the war in November of 1918, all sense of urgency surrounding the laboratory was lost and from this point onward the laboratory, as far as the board was concerned, was essentially a dead issue and they never again actively pursued its development. The lack of a laboratory, however, provided grounds for numerous complaints and the failure of the laboratory project became a casus belli for members, particularly Sprague, in their increasingly antagonist relations with the navy and Navy Secretary Daniels. On several occasions, the failure of the laboratory project was the gauntlet thrown down by one or another member in a threat to resign the board, including one incident in March of 1919, in which the entire board (with the exception of Edison) threatened to resign over the issue (these matters will be explored in greater depth in the next section). A few months after the failure of the second proposal, Secretary Robbins grimly pointed out that while they, the board, still didn't have a laboratory, the National Research Council had obtained the use of the laboratories and machine shop of the Carnegie Institute and was now actively pursuing research. The board had simply lost out on the issue.[90]

Sprague seems to have taken the rejection of the laboratory plan somewhat harder than the rejections of his submarine and mine work. Although he faulted the navy's foot dragging and lack of support and enthusiasm as well as Navy Secretary Daniels personally, he clearly apportioned the lion's share of the blame to Edison, and he made a point of saying so in numerous letters. Sprague came to feel so strongly on the issue that he tendered his resignation from the board in a harshly worded letter to Navy Secretary Daniels in January 1920. In the letter, he blamed both Edison and Daniels not just for the failure of the laboratory proposal, but for the overall loss of significance and dignity that the board and its members had suffered. He went to great lengths, however, to praise his other colleagues on the board and even Daniels himself for the cordial and friendly relations that they had always enjoyed.[91] This move was received with approval and respect by many of his board colleagues who expressed hope that this would finally provoke some action. It also prompted one worried response that Sprague should withdraw the letter and an offer to remove it from the files before Secretary Daniels had a chance to read it.[92]

Sprague was persuaded to withdraw his resignation a short time later by a letter from Navy Secretary Daniels, in which Daniels promised Sprague that the plans were finally underway, that construction would soon begin, and that this would be "an opportunity . . . to aid in a constructive work that will last for all time." The letter also included a hand-written apology for the delay.[93] This wasn't the end, and Sprague had to write more angry letters and threaten to resign once more before the laboratory was finally constructed.[94]

The laboratory finally was completed in the fall of 1922, at Bellevue, as the board had recommended in 1918. Sprague was sent an aerial photograph of the newly completed laboratory, dated October 1922, nearly seven years after the board had begun advocating its construction. This probably struck Sprague as far too little and too late.

THE SIGNIFICANCE OF THE NAVAL CONSULTING BOARD

Although he may never have said so directly, it is clear from Sprague's letters that he considered the board to have been a failure. One might be tempted to attribute his attitude to personal bitterness over the failures of his various endeavors, but the sentiment seems to have been widely shared by his colleagues. As early as 1915 there were complaints that the navy was already ignoring the board, and by 1917 several of the members were dismissing it as merely a political tool of the navy, Navy Secretary Daniels, and President Wilson.[95] This culminated in the mass resignation threat which was only narrowly staved off by a direct appeal from Daniels.[96]

Sprague clearly shared these frustrations with his colleagues, and they provided the background for his personal frustrations with the rejection or failure of his individual projects. On top of all this, he had difficult relations with board president Edison. Tensions between Edison and Sprague are apparent from the outset and may reflect a degree of ongoing antagonism between the two derived from earlier interactions as well as legitimate conflicts that arose as part of their board duties.

Sprague had already engaged in some sparring over public perceptions of Edison's role on the board.[97] Beyond these individual "spats," Sprague and the other board members had some substantive issues with Edison. On a number of occasions, Edison simply declared his solution to a problem superior and the matter was closed (submarine batteries and hydrogen absorption systems are two examples).[98] As the controversy over the lab developed through the war and into the post-war years, Edison was seen as increasingly problematic by, and became increasingly isolated from, the other board members—much of his later communication with the board's other members was through intermediaries. In a letter to board historian Scott, Sprague had been harshly and pointedly critical of Edison, and had laid the lion's share of the blame for the board's failings at his feet.[99] In a later letter, written in regard to a second edition of the board history, Sprague complains to Scott that his (Sprague's) comments, particularly about Edison, had been ignored in the first edition.[100]

In the end, it is probably fair to say that the Naval Consulting Board was a failure, at least in terms of the members' visions for it. After reviewing thousands of submission, the board only saw one through to production, although a number of other inventions eventually were patented as a result of research sponsored or inspired by the board. The Naval Laboratory never came to fruition in the form the board had envisioned, and to the extent that it was realized, it was far too late to make any difference in the war effort. Finally, the board was never taken seriously by the navy it was intended to serve. One of the most persistent complaints of Sprague and other board members was the lack of consideration shown the board by the navy and its representatives.

Naval Research Laboratory
Bellevue, D.C.
October-1922.

Despite strong disagreements on the location of its site and difficulty in getting sufficient funding, the Naval Research Laboratory was finally completed in 1922 at Bellevue on the east bank of the Potomac River in Washington, D.C. By this time, the Great War was long over. But the Naval Research Laboratory continues today to make distinguished contributions to military science and technology, and to their transition to practical applications, with a distinguished faculty that includes winners of one Nobel Prize and numerous other awards for scientific achievement. *Frank Sprague Papers, Manuscripts and Archives Division, The New York Public Library, Astor, Lenox and Tilden Foundations.*

Taking a broader view of the board and its mission, however, some positive contributions can be identified. First, given the disparity in industrial preparedness between the United States and Germany at the outset of the war, the significance of the Industrial Preparedness Campaign cannot be underestimated. The United States probably could not have mounted to a war footing as rapidly as it did without this broad survey of its industrial capacity.

Second, the board did serve a "political" mission as its members had complained: while they were frustrated by the tedious task of reviewing the public's outflow of inventions, the board provided an official venue for the public to "make a contribution" to the war effort, which served the navy's interests in several regards. The premise of the board played very well to Americans' sense of self as a "naturally inventive people"; anyone could make a contribution that could help to end the war. It also presupposed the notion, again appealing to Americans' sense of self, that the war, despite Sprague's charges to the contrary, would be won by technological innovation. Both of these points helped to distract America from the alarming fact that Germany really was better situated, both technologically and industrially, than the United States at the outset of the war.

Finally, while almost no innovations to naval warfare emerged directly from the board's work during the First World War, many of the technologies that they explored later did become important tools, techniques, or tactics in later wars, notably the Second World War. PT boats, submarine detection systems, a variety of naval ordnance, and naval tactics all benefited from important theoretical and practical testing and evaluation by the board's various committees and members. While the board did not shepherd these developments through to fruition, it helped to lay some of the groundwork for later developments.

THE END OF THE BOARD

The Naval Consulting Board began in a flurry of activity that carried it through the prewar and war years. With the end of hostilities in 1918, much of the urgency that had surrounded the board was understandably dissipated and the board gradually wound down its activities. During the post-war years there were increasingly frequent calls by various board members for the board to be disbanded or for its members to resign. Meetings and other activities became decreasingly frequent to the point where, in one letter, Sprague described the board as "moribund" and "never meeting."[101]

There is a sense of poignancy in the board's final communications: invitations to the annual dinners, a reminder that board members had promised "as long as they were alive" to attend an annual Armistice Day dinner, and one of the very last, a letter informing the board members that tickets to the Army–Navy game would no longer be provided free of charge, but could be purchased for $3 apiece.

OPPOSITE Frank Julian Sprague (1857–1934) photographed circa 1921. *Frank Sprague Papers, Manuscripts and Archives Division, The New York Public Library, Astor, Lenox and Tilden Foundations.*

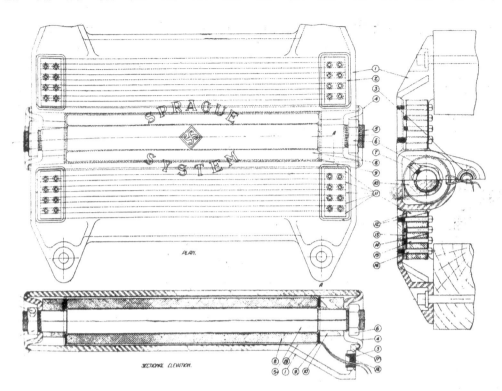

PLAN.

SECTIONAL ELEVATION.

The Permanent Magnet Bars and the Electro Magnet Form the Track Unit

Extended tests of the automatic train control system developed by Frank Sprague were made on a New York Central line between Ossining and Tarrytown, New York, in 1922 and early 1923. This drawing shows a typical permanent magnet bar with two groups of bars and an electromagnetic core between them. These pieces were enclosed in a sealed manganese steel casing for protection and bolted to the track ties.   Railway Signaling, *January 1925. Courtesy Simmons-Boardman Publishing Corp.*

9

SPRAGUE AND RAILROAD SAFETY

By the beginning of the twentieth century the United States had emerged as the world's greatest industrial power, and it was supported by one of the largest railway systems in the world, reaching virtually every corner of the continent. In the period from the time of the Civil War to the end of the century, the railroads transported an American manufacturing output that had increased fivefold. Coal production had grown by ten times, with an annual production that reached 270 million tons annually by 1900. Steelmaking had grown into a giant industry that had produced more than 10 million long tons of steel in 1900. The output of American farms had almost tripled between the Civil War and 1900, and the products of the nation's forests had more than tripled to reach more than 35 million board feet annually in 1900.

By this time American railroads had grown into an industry of some 259,000 miles of track that were transporting an annual traffic of almost 142 billion freight ton-miles and more than 16 billion passenger-miles. In just the decade from 1890 to 1900 freight traffic had nearly doubled, and passenger travel had grown by a third. As railroad traffic continued to grow in the new century, the railroads were meeting the demand with new and heavier locomotives, larger cars, and expanded and strengthened tracks and bridges.[1]

But this steadily growing business and earnings of American railroads was accompanied by a very much less desirable one, a rapidly growing record of deaths, injuries, and property losses from railroad accidents. In 1900, for example, Interstate Commerce Commission data showed a total of 7,865 fatalities on American railroads.[2] The majority of these were by trespassers, but the fatalities to passengers and railroad employees—2,799 in 1900—were staggering enough. The causes of train accidents were roughly equally divided between derailments and collisions, with each averaging more than a thousand annually in 1900.[3] By this time there was a growing public concern about the railroad accident problem. In 1901 Congress passed the Accident Report Act, requiring the listing of all accidents over $150 in damage or which caused a casualty. The 1910 Accident Reports Act gave the Interstate Commerce Commission authority to investigate train accidents. These and other measures gave the Commission an increasing authority over the safety practices of the railroads.

Much of the attention given to the problem of train collision was focused on signal protection. American railroads had historically used the telegraph and train order system of train protection. This worked reasonably well in light traffic, but the history of the railroads is filled with stories of misunderstood orders, overlooked or forgotten meeting points, or a failure to provide adequate flagging that led to a collision. As traffic increased, railroads began to use manual block signals to provide better train protection, and these in turn were gradually replaced with automatic block signals beginning with their first development in 1883. But even as late as 1908 the majority of railroads still operated under train order control. Of some 151,500 miles of line carrying passenger trains at the beginning of 1908, only about 58,700 miles of these were operated under block signals, with only about 10,800 of those under automatic block signals.[4]

In 1904 the Interstate Commerce Commission reported that the great loss of life in train collisions was preventable through the use of block signals, and the Commission began to ask Congress for authority to study the needs of block signals and to require their installation. An Interstate Commerce Commission Block Signal and Train-Control Board appointed in 1907 was chaired by Professor Mortimer E. Cooley, Dean of the Department of Engineering at the University of Michigan, and included three other members experienced in railroad operations. The board was appointed to investigate and test devices developed for automatic train stopping, as well as all types of block signals. In its first annual report, the board commented:

> As the block system proper is now in general use and information concerning it can be obtained with comparative ease the board has devoted its attention mainly to the question of automatic stops. In considering the desirability of such devices on railroads generally one of the first problems to be solved is

the reliability of the apparatus when its operation is interfered with by snow, ice or accidental disturbances or obstructions.

The board gave its first priority to devices which were already in use, with second priority being given to a large number of inventions and alleged inventions which had not yet been developed or installed. It found little of value in most of them. "As to the second class, all but a few of the devices that have been presented for consideration are from inventors who have little knowledge of the present state of the art of signaling and who appear to have taken little pains to secure the counsel of men experienced in railroad operation."[5] In nearly all of the 55 cases which had been passed upon by the board, it had been unanimously decided that there was not sufficient merit to warrant further attention. The board's findings were discussed in annual reports until June 1912, when it was disbanded for lack of further funding.

In 1914 a committee of the Railway Signal Association was requested by the American Railway Association to draw up a set of requirements for the installation of train control equipment. These were adopted by the commission on May 20, 1914. Later, in 1919, the same requirements were also adopted by the Interstate Commerce Commission's Safety Bureau. As adopted, the requisites established:

> An installation so arranged that its operation will automatically result in either one or the other or both of the following conditions:
> First—The application of the brakes until the train had been brought to a stop.
> Second—The application of the brakes when the speed of the train exceeds a prescribed rate and continues until the speed has been reduced to a predetermined rate.

These overall statements were followed by thirteen detailed requisites concerning details of the equipment's function and installation, and a description of four permissible adjuncts to train control equipment, which included cab signals, detonating signal apparatus, speed indication, and recording devices.[6]

The work of the board was later taken up by the Interstate Commerce Commission's Bureau of Safety, reporting on plans submitted to it and making tests. The idea of an automatic train stop that would be activated whenever an engineman had failed to comply with a signal remained a popular idea with inventors, and the Bureau of Safety reported that some 400 devices had been submitted by inventors, but only a few of the devices were of sufficient merit for further development or testing. After more than a decade of studies and testing however, the Interstate Commerce Commission had little to show for it, and the development of an automatic train stop that would work in the demanding railroad setting remained still only an idea.

There had been several early trials with simple automatic stopping devices, but none of these were very suitable for a railroad setting. As early as 1889 the Pennsylvania Railroad at its Gallitzin Tunnel on the Alleghenies summit had installed a Union Switch & Signal Company system of glass tubes standing above the roof of the cab, which would be broken by a projecting arm from a signal mast if a stop had been ordered. But the tubes were so frequently broken by icicles during the winter, requiring unnecessary stops, that the system was quickly abandoned.[7] Similar equipments were tried by rapid transit systems with much greater success. The Boston Elevated subway installed a system in 1901 that used a raised arm or trip to release the air and set the brake if a train passed a restricted signal. Similar installations were put in place in New York's Interborough Rapid Transit and the Hudson & Manhattan Railway under the Hudson River.[8] By 1922, *Railway Age* had reported that about 3,700 electric or electro-pneumatic automatic train stops were in use on New York's Interborough, Brooklyn Rapid Transit, and Hudson & Manhattan lines, as well as a few on the Pennsylvania's Hudson and East river tunnels and drawbridges.[9]

With Frank Sprague's consulting work on the electrification of the New York Central railroad largely completed by the end of 1906, he turned his attention to the difficult technical task of making railway operation safer in general by automatically forcing obedience to wayside signals without undue interference with the proper control of the train by the engineer. Sprague formed the Sprague Safety Control & Signal Corp. in 1907, and his experimental study work went forward over the next several years. By 1913 Sprague had developed his automatic train control system to the level that he could describe the new system in a September 9 letter to Charles A. Prouty, Chairman of the Interstate Commerce Commission, offering to make preliminary stand tests of the equipment.

Nothing happened at the Commission at the time, but Sprague's work continued. The system had undergone complete shop tests extending over three years, and road and track tests on General Electric tracks in Schenectady extending over a year. Patent attorney Lawrence Langner recalled the tests in Schenectady:

> I well remember one bleak morning in Schenectady riding with you on an
> electric train equipped with your now famous safety device. The train pur-

posely ran past the signals, and a ghostly hand clutched at the train and brought it gently to a standstill.

I prophesy that long after you and I have passed on to the place where mechanism will be of no use to us the Sprague Automatic Train Control will be safeguarding the lives of untold millions.[10]

By 1915 the work had been advanced to the point at which it had undergone complete shop tests and the Commission's Henry C. Hall could be provided with a detailed description of the Sprague Safety Control System. In a June 23, 1915, letter Sprague Safety Control's O. B. Wilcox requested that the completed installation be tested for the Commission's Block Signal and Train Control Board, with these assurances:

> we feel justified in asserting that any required equipment in trunk railway line work may be undertaken, under conditions of guaranteed non-interference with existing signal and braking systems; and that the System when so introduced will constitute a dependable safeguard against rear-end collisions on single and double tracks, and head-on collisions on single tracks, while at the same time not interfering unnecessarily with the regular operation of trains and the maintenance of schedules.[11]

At about the same time, Sprague also filed a patent application for his work in August 1914, but it would be another 16 years of disagreement between Sprague and the Patent Office before the Sprague automatic train control patent was finally approved in April 1930.

The Sprague Safety Control System, said O. B. Wilcox,

> involves no impact of a mechanical trigger or third rail or ramp contact device, is not influenced by weather conditions, and is applicable to all railways, whatever the motive power, and in combination with existing standard train and car air-braking and track signaling systems, without involving in any way changes in either the track or signal circuits or traffic rails.
>
> The System adds greatly to the safety of life and property by promoting vigilance, correcting the failure of engine-men to observe signals and speed regulations, and providing an automatic report of violation or rules without, however, imposing any objectionable or unnecessary limitations upon the control of the train by the engine-man.
>
> It permits of an increase in schedule speeds and traffic handling capacity through enabling trains to be safely operated in shorter time intervals and at increased average speed.
>
> It promotes greater economy in operation by reducing casualties and loss and avoidable delays, and of equal importance, by reducing the "emer-

A magnetic receiver unit was located on the locomotive's tender, consisting of two flat iron collector plates riveted into a locked non-magnetic casting, and brake applications and reset magnets which provided a magnetic flux picked up from the receiver plates. The first photograph shows the sealed magnetic receiver, while an installed unit is shown in the other.   Journal of The Franklin Institute, *July–December 1922, from the Historical and Interpretive Collections of the Franklin Institute, Philadephia, Pa.*

gency" brake applications and "rough" handling of equipment. Automatic applications of the brakes with this System are measured in accordance with the requirements of each character of train, viz: the automatic application is made in the same manner as a competent and alert engine-driver would himself make it under the existing conditions as they happened to be at the time of the automatic application.

These photographs show an air valve assembly mounted under the cab of the test locomotive (*right*), and the main engineer's valve and signal lights in the engine cab (*below*).   Journal of The Franklin Institute, *July–December 1922, from the Historical and Interpretive Collections of the Franklin Institute,* Philadephia, Pa.

It forestalls heavy capital expenditures for additional main and passing tracks, improved train dispatching, roadway, and other facilities that an automatic train speed control system makes unnecessary. It does not involve the discarding of existing investment, and encroaches in no way upon the clearance limits of roadbed or equipment.[12]

The Sprague system of automatic train control employed what was called an intermittent magnetic induction type, using permanent electromagnetically controlled track magnets operating on the normal danger principle. In addition to automatic stop, the Sprague system also included graduated speed control and cab signal features.

Trackway elements of the Sprague system provided brake application magnets, an assembly of permanent magnets with an electromagnet of opposite polarity which could be used to "clear" a train under proper signal conditions. These were sealed in a manganese steel casing for protection and bolted to the track ties, with the top surface of the magnet assembly located 4 to 4½ inches below the rail height. The fixed magnets were positioned at sufficient space between the magnets and the location of the home or distant points so that a train could be safely and automatically braked to a halt whenever the line was set against train movement. If a track was clear, the magnets would clear a train through a signal, while an inadvertent passage through a restrictive signal of a speed restriction would initiate the automatic brake application.

Magnetic receivers were mounted on the locomotive's tender, suspended 4 to 5 inches above the top of the rails. These consisted of two flat iron collector plates riveted into a locked, nonmagnetic casting and brake applications and reset magnets which provided a magnetic flux picked up from the receiver impulses which were then transformed into braking action. A speed-brake governor and control was installed on the locomotive's pony truck and could control speeds as required for tangents and curves, or secondary or emergency brake control. Other equipment included a variety of application- and reset-magnets, control relays, air brake valves, light indications and audible alarm, and a recorder.[13]

The system was developed to meet several insistent requirements of the railroads. There was a demand "that there should be no unnecessary interference with the engineman so long as he was safely attending to his business, and that there should be no unnecessary automatic stopping of trains." A recommendation of the automatic train control committee of the United States Railroad Administration made speed control a dominant feature. The advice and constant urging of the principal signal companies was expressed in the slogan: "Keep trains moving."[14]

With the United States's entry into the First World War in April 1917, followed by the federal takeover of American railroads late in December 1917, little progress was being made with the development of automatic train control systems. During this time, Frank Sprague was heavily involved in his work with the Naval Consulting Board. The Interstate Commerce Commission's Bureau of Safety and Frank Sprague periodically debated the details of the Sprague train control system, and he became a frequent writer to the Interstate Commerce Commission, members of Congress, and publications, pointing out recent examples of railroad collisions which could have been avoided through the use of the already proven Sprague system.

On July 11, 1918, for example, Sprague wrote to William McAdoo, director general of the railways, noting the high loss of life on two recent collisions on the Michigan Central and the Nashville, Chattanooga & St. Louis, and pointing out how the Sprague Safety Control & Signal Corporation had invested a quarter of a million dollars in developing and testing his system of automatic train control "for the purpose of protecting the lives of passengers and employees, and the property of railroads, from accidents which result from failure of enginemen to observe, to properly interpret, and to obey signal indications, from any cause whatever."[15]

A year later, Sprague wrote to Senator Albert B. Cummins, Chairman of the Senate Committee on Interstate Commerce, expressing his concern over the long inability to get the railroads to introduce the now well-developed systems for automatic train control, and expressing the hope that the government would gain the ability to insist that safety devices be adopted by the railroads as they returned to private ownership. Various forms of automatic stops or train control had been developed by inventors at an expense of millions of dollars—in Sprague's own case his costs had climbed to more than $350,000—in response to the recommendations of the Interstate Commerce Commission "and to the increasing demand on the part of the general public for some means to prevent the recurrence of the numerous wrecks due to man failures."[16]

Just a few months later, Sprague was back with another letter, this time to the editors of *Railway Age* in their February 6, 1920, issue. Calling his piece "The Need for Automatic Train Control," Sprague cited passenger train collisions over the previous two months which had killed a total of 43 people. "These recent news headlines telling the all too familiar story of butting (end-to-end) collisions, are but typical of that class of accidents which are not only preventable but the continuation of which is absolutely inexcusable."

Sprague went on to criticize much of the railway industry for its opposition to the adoption of automatic train control, which he already considered a proven technology. A. P. Thom, who represented the Association of Railway Executives before a recent hearing of House and Senate conferees, opposed the inclusion of a mandatory provision in the Esch[17] bill concerning the use of automatic train control. Thom quoted Chicago & Great Western president S. M. Felton as the best-qualified person in the country to give an opinion on the practicality of automatic train stops. Felton declared: "It as an utterly impractical scheme; it removes the responsibility of handling the train from the enginemen, and we would have much more serious accidents than we now have. An engineer would not have to be awake, alert and on the job all the time, because we would soon depend on the automatic device to protect him."[18] Sprague highlighted recent innovations as a basis for his argument in favor of this new step, too:

> The same profession of impossibility of engineering accomplishment has steadily opposed every radical improvement in railway equipment and operation; in the face of which the air brake has been developed, the car stove has been discarded, the automatic car coupler has been prescribed, the automatic block signal system has been installed and steel cars are taking the place of wooden ones; and grade crossings, despite the prophesy of bankruptcy by the head of one of the most important roads entering Chicago, have been abolished in that city.[19]

But finally there were some reasons for optimism. Among other things, Section 26 of the Transportation Act of 1920, passed on February 28, 1920, authorized the Interstate Commerce Commission to require automatic train control or "other safety devices," and the railroads returned to private ownership and operation on March 1, 1920.

There were still other letters from Frank Sprague, urging prompt installation of automatic train control systems. In letters of June 8 and July 1, 1920, to W. O. Borland, Chief Bureau of Safety, Interstate Commerce Commission, Sprague cited recent train collisions on the New York Central near Schenectady and Buffalo, and others on the Pere Marquette, Delaware & Hudson, and Boston & Albany, and pointed out the added responsibility to the Interstate Commerce Commission now that it had its new authority to enforce the safety recommendations that it made. Using some of the funding made available to the railroads from Congress under the reversion of the railroads to private op-

Our Compliments for the Season

SPRAGUE

SAFETY CONTROL & SIGNAL CORPORATION

421 Canal St.
New York

310 Michigan Ave.
So. Chicago

This rather bizarre greeting to railroad men went out in the advertising pages of *Railway Age* in its January 10, 1925, issue. *Courtesy Simmons-Boardman Publishing Corp.*

eration, Sprague suggested that it could be used for testing of automatic train control systems to identify those that would meet the Commission's standards for use on the railroads.[20]

As the prospects improved for the development of automatic train control, Frank Sprague began to look at a possible sale of his Sprague Safety Control & Signaling Corp., much in the same way he had sold off such earlier companies as Sprague Electric Railway & Motor Co., Sprague Electric Elevator Co., and Sprague Electric Co.

Following a meeting early in 1920 with H. H. Westinghouse (George Westinghouse's younger brother), board chairman of the Westinghouse Air Brake Company, and A. L. Humphrey, president of Westinghouse, Sprague had submitted a proposal to Humphrey for consolidation of the Sprague Safety Control & Signaling Corp. with Westinghouse Air Brake and the affiliated Union Switch & Signal Co. Outlining his belief that automatic train control would soon become a subject of active commercial exploitation, Sprague proposed that the Sprague firm and the Westinghouse companies be brought together to expand the Sprague project with a subscription of stock and board representation, working toward some later form of affiliation or consolidation. While his firm had adequate financial resources, Sprague recognized that the resources of a much larger railway braking and signal company would likely be of great help.[21]

But the prospective connection with Westinghouse never went ahead, very likely because Westinghouse's Union Switch & Signal Co. was already at work developing its own automatic train control system, and the Sprague Safety Control & Signaling Co. would continue as a separate company. Sprague's eldest son, Desmond, who had worked with his father in the Naval Consulting Board during the war, also joined his father in the safety and signaling company at this time and remained with it until it was closed in 1933. Other key men working with Sprague on the train control system included Harold B. Cockerline, M. Knab, W. L. Hauck, and F. M. Shannon.

Sprague's idea of a central fund from Interstate Commerce Committee resources never did materialize. But in August 1920 a committee made up of members of the New York Public Service Commission and New York Central staff was appointed to develop a plan for testing multiple automatic train safety systems, and the following March proposed a location of a two- or three-block section of the New York Central's Mohawk division between Schenectady and Hoffmans.[22] This plan, too, failed to materialize, at least as visualized, but by late August the New York Central had agreed with the American Railway Association to make available a single track line covering six blocks in a 6½-mile-long section of track on the four-track main line between Ossining and Tarrytown, New York, on the Central's main line north of New York. Unlike the previous arrangement, which was to be made available to multiple automatic train control suppliers, the Ossining track was to be made available only for tests of the Sprague system. The railroad was to furnish the expense of installing the track and line control circuits, as well as that of installing the track apparatus and engine equipment, and would assume the expense of engine and train crews, fuel and other supplies. Sprague was to furnish and maintain all track magnets and engine equipment of the Sprague system, and supervise the installation and maintenance of the apparatus.[23]

Trials were conducted on the test track from May 1, 1922, to January 31, 1923, using New York Central K-3 Pacific No. 3345 and a single coach used for test equipment. Typically, these were conducted seven hours a day, six days a week. Extensive runs were made between May 1 and November 20, 1922, before the inspectors of the Commission began their final review, which then ran through January 31, 1923. The trials, the first of their kind ever made, drew frequent visitors, railway signal men, and others to both the test site and Sprague's 421 Canal Street shop in New York. The final tests for the Sprague system included 145 trips and 724 trials, with each movement over a track magnet being counted as one operation test. From a total of 2,177 operations all but three magnet tests operated satisfactorily, with one safe failure, one false clear operation, and one of unsatisfactory track magnet operation. A report on the Sprague automatic train control system, published in *Railway Age*, provided a generally positive outcome of the testing period.

The trials showed that there was some undesirable interference from current from the New York Central's third rail electrification, which could be overcome by the use of suitable means of protection, and that the fullest measure of protection would require the insertion of certain insulated rail joints, which had not been done during the tests.

In almost every respect, the Sprague equipment worked well. While there was severe weather from December 5 through the end of the testing period, and the installed permanent magnets were buried in snow and ice, there were no ill effects on their operation, and the electromagnets used for the reset function could be relied upon at all times to provide maximum protection. The tests indicated that the Sprague automatic train control system could be installed and interconnected with existing signal systems, which merely showed a signal indication, and that it worked well with them.

Further testing would be required to determine if the system would work well under the varying needs and conditions of everyday freight service, or if there might be any possible effects for its use outside of electrification territory.

Overall, the Interstate Commerce Commission test men found that the Sprague components of the locomotive apparatus, speed governor, relay assembly, pilot valve magnets, the visible cab indicators, the receiver, and the pneumatic apparatus were all well designed and constructed, and capable of providing a reliable operation.

> While, as a whole, the observations and service tests made are not considered conclusive, it has been determined that a magnetic impulse can be transmitted in electrified territory from permanent track magnets to the locomotive regardless of speed, oscillation, or weather conditions, and that such impulse will actuate the locomotive apparatus to provide automatic brake application in a practical manner. In view of the results obtained under the conditions surrounding these tests, the conclusion is warranted that this device has such inherent merit that a more extensive installation should be made where the real value of the system can be more fully demonstrated.[24]

An even more emphatic testimony of the value of the Sprague automatic train control system came from a letter of December 24, 1923, from John F. Kelly, Senior Mechanical Engineer, Interstate Commerce Commission, to C. O. Bradshaw, Assistant General Manager of the Chicago, Milwaukee & St. Paul Railway, who doubtless had asked for advice on automatic train control systems. Kelly had put his 35 years in the railroad engineering profession into developing the most exhaustive possible testing program for the Sprague system on the New York Central test track. He was particularly impressed with the performance of the air braking system.

> I can say without qualification that the air brake mechanism of the Sprague system of automatic train control is far superior to any other thus far submitted to the Commission for examination; and taking the air brake condition as we find it to exist at present it is the only one capable of operating the brake safely on long freight trains.

Kelly summarized these thoughts from the tests of the Sprague equipment:

1. The Sprague train control system is absolutely reliable.
2. Failure, if it occurs, will be on the safe side.
3. Safe failures will be extremely rare.
4. It fulfills all of the requirements of the Commission's order specifying what train control should do, which requirements are based on careful

examination of the accident investigation records of the Commission covering a long period of years, and it covered all that is necessary or useful.

5. It will not interfere with the orderly operation of the train or annoy the engineman.
6. It will afford absolutely reliable adequate train protection.
7. It will not interfere with the use of the locomotive outside of equipped territory.

It is a perfected system in all its elements that has demonstrated its ability to meet reliably and perfectly all the essentials and all the practical refinements which a train order system should meet, and its durability and freedom from defect has also been rigorously demonstrated.[25]

The Interstate Commerce Commission had gained authority via the 1920 Transportation Act to require the installation of automatic train control or other automatic block signals, and early in 1922 the Commission finally moved ahead to mandate the development of the automatic train control systems that the Commission and such interested safety proponents as Frank Sprague had been advocating for so long. The Interstate Commerce Commission order No. 13413, published on January 13, 1922, ordered 49 railroad companies to install automatic train control devices on specific congested sections of the railroads to be operated in connection with all road locomotives. Particular equipment to be used was not specified by the Commission, but the completed installations had to be inspected and approved as meeting the Commission's specifications and requirements. The work was to be completed by July 1, 1924. This would require the installation of devices on about 6,126 miles of track, and the outfitting of an estimated 5,525 locomotives with the equipment, the cost of which would be anywhere from $100 million to $200 million, guessed the *Christian Science Monitor,* while the American Railway Association could only estimate that it would "aggregate many millions."[26]

In issuing the order, the Commission noted a wreck on the Philadelphia & Reading near Philadelphia the previous month in which 33 persons had been killed, and which the Commission maintained, "would not have occurred" had there been adequate automatic control devices. Also cited was a recent rear-end collision of two Pennsylvania Railroad trains near Manhattan Transfer which "doubtless could have been prevented had the automatic train control system, in use from the Pennsylvania Terminal, New York City, to the Hackensack River, been extended to the Manhattan Transfer, a distance of some two miles."

"Our investigations have shown that automatic train control has long since passed the experimental stage," reported the Commission, "in fact, no safety device such as the automatic coupler, the airbrake and the automatic block signal were perfected to as high a degree as the automatic train control before they were either ordered installed or were voluntarily adopted."[27]

Prospective automatic train control systems were classified into several types of installation. Intermittent control equipment transmitted a signal from the track to the train only at certain fixed points in the track, usually at or near signals, with the indication continuing until the next indication point was reached, at which time it could be continued or changed. In the continuous classification of control equipment the indication from the track to train would be continuous, usually from the rail, and the indication could change at any time. Contact between the track and the train for either intermittent or continuous systems could be either contact, using some type of mechanical or electrically controlled trip, either on the ground or overhead, or a non-contact trip using induction with permanent magnets or inert roadside elements.[28]

There was no shortage of prospective suppliers of automatic train control equipment. A *Railway Age* article in October 1922 listed no less than 36 different suppliers of automatic stop or train control systems that had already been put to use or tested.

Several elevated or subway rapid transit systems, as well as several main line railroads had already installed some type of automatic train stop or train control system well before the Interstate Commerce Commission had ordered any such installations. In addition to rapid transit railways, the Pennsylvania Railroad had been operating an automatic stop system under the Hudson and East river tunnels and its electrified track to Manhattan Transfer since 1911. In 1913 the Chicago & Eastern Illinois installed the Miller Train Control Company automatic stop system over a 107-mile, double track main line section between Chicago and Danville, Illinois. During 1916 and 1917 the Chesapeake & Ohio installed a 21-mile section of line between Charlottesville and Gordonsville, Virginia, using the American Train Control Company system. Another early main line user of automatic train control was the Chicago, Rock Island & Pacific, which installed a 22-mile section of line between Blue Island and Joliet, Illinois, in 1919, using the Regan Safety Devices Company intermittent electrical contact system.

These early users of automatic train control were typically positive about its use, finding that it did increase safety, and also contributed to more efficient train operation, permitting an increased number of trains. But these supporters were in the minority, and most of the railroad companies were strongly opposed to the Interstate Commerce Committee's mandated installation of automatic train control. Soon after the Commission's order was published, the American Railway Association urged that the order not be carried out, stating that automatic train control had not reached a point of development justifying an order for its extensive installation, that further tests were needed to determine the most practical and reliable types that should be used, that the carriers were already making every reasonable effort to cooperate with the Commission, and that the proposed order would make necessary far more and expensive installations than were warranted by the present state of automatic train control development.[29]

Still other objections were made by members of the individual railroad companies. There were a few who simply opposed using an automatic train control system, arguing that using an automatic device would simply transfer from the engineman the care and attention that was required from him to the equipment, with no improvement in safety. But much more common were the objections of railroad men, who argued simply that with the limited safety funding that they had available, they could make greater improvements in safety by further extending automatic block signals or by eliminating highway grade crossings, which by the 1920s had become a rapidly growing safety problem for the railroads.

Commission hearings held in April 1922, as well as separate appeals from the railroads, brought a number of changes and time extensions. The railroads were hard at work on the 1922 order, when the Commission issued a new train control order, published on January 14, 1924, ordering 45 additional roads to complete automatic train control installations, while 47 of the roads listed in the original 1922 order were now required to add train stops or train control for an additional passenger locomotive division. All of the new installations were to be complete by February 1, 1926.[30]

"To many, this last order came somewhat like a bolt out of the blue sky," wrote *Railway Age*. A petition signed by 88 Class I railroads asked that the new train order be annulled, and that the deadline for completion of the installations called for in the original order be extended another year, to January 1, 1926. The Commission denied all of these requests except for allowing a hearing for the new lines added by the second order. Following hearings in the spring of 1924, the second automatic train control order was modified to cancel the requirements for new installations from the 45 companies added by the second order, while the order requiring additional installations from the original list of railroads remained. The original order was also modified to permit the use of a "permissive feature," under which an alert engineman could forestall automatic setting of brakes at a restrictive block setting. Plans were also made for further joint testing of automatic train control equipment conducted jointly by the Commission and the carriers. Commissioner John J. Esch, who had been a long-time proponent of mandatory automatic train control equipment, strongly dissented from the Commission's relaxation of its orders. In any case, the second order of 1924, as modified, would turn out to be the Commission's last order for mandatory automatic train control equipment.

On November 27, 1928, the Commission issued a decision that it would not order the installation of any further train control at that time. The commissioners expressed the opinion that "vigorous efforts to provide adequate protection against accidents due to grade crossings, derailments and collisions in territory not protected by block signals, failure of wooden bridges and trestles and use of wooden passenger cars will afford a far greater measure of safety than requiring by order special efforts to extend train control installations."[31]

While all of the appeals and hearings of the Commission's automatic train control orders were being held, the railroad companies were beginning to get the signaling work done. By early 1924 the early beginners of automatic train control, the Chicago & Eastern Illinois, the Rock Island, and the Chesapeake & Ohio, had already received the Commission's approval of their systems, or were in the process of getting approval. Some 15 railroads were already under contract for the installation of systems, while another 19 were conducting road tests of equipment.[32] By the end of 1926 more than 11,145 miles of tracks with about 4,550 locomotives were in service, representing 85 percent of the track miles and locomotives required under the first order, while 62 percent of the track miles and 45 percent of the locomotives required for the second order were complete.[33] By July 1, 1928, the automatic train control installations ordered by the Interstate Commerce Commission were almost fully complete, with 11,185 miles of road and some 8,127 locomotives in operation.[34] Costs for the work were reported at almost $12.9 million on May 31, 1927, at which time almost all of the work under the two orders was complete.[35]

By and large, the railroad users of automatic train control were well satisfied with its use. The equipment worked well, and the train control aspects of the system proved to be a significant aid in moving traffic more effectively and promptly. While the effects of train control on train safety were hard to quantify, in November 1930 the Commission noted that with 9,000 locomotives and 21,000 miles of road equipped during the previous four to six years, not a single disastrous collision had occurred on any of the controlled line. Several railroads even went to the extreme of advertising their use of automatic train control.[36] An advertisement from the Chicago & North Western in 1927 extolled the increased safety of train operation with the installation of automatic train control between Omaha and Clinton, Iowa, with the new improvement soon going all the way to Chicago. The Milwaukee Road, in a booklet about the newly equipped *Pioneer Limited,* described the new train control equipment: "One of the most interesting features is the automatic train stop, which device protects the passenger against accidents due to broken rails or failure of engineers to observe signals."[37]

With total expenditures approaching $13 million, the automatic train control project had represented a significant venture for the railroads and their suppliers. But for the specialized automatic train control suppliers who had developed the new train control systems, it was anything but. Of the dozens of potential suppliers who had developed proposals, only four of these specialized suppliers were ever awarded contracts, and few of these were of much significance. Instead, the two giants of railway signaling—Union Switch & Signal Co. and General Railway Signal Company—had developed new automatic train control equipment for the new market, and the railroad companies awarded the overwhelming share of business to their traditional supplier manufacturers. Fully 75 percent of the contracts and more than 80 percent of the total train control mileage went to either Union Switch & Signal or General Railway Signal. Union and General developed equipment for use on both continuous and intermittent signaling systems, while the Sprague system, for example, could be used only on an intermittent signal system. Almost half (about 45 percent) of the automatic train control systems carried out under the ICC orders were designed for continuous signaling systems.

Frank Sprague was originally scheduled to develop the train control system for the New York Central Railroad's Mohawk Division between Albany and Syracuse, which included some 582 miles of track and 320 locomotives to be equipped, but this work was later shifted to General Railway Signal. In the end, Sprague had only four contracts, to the Chicago, Burlington & Quincy, the Northern Pacific, the Great Northern, and the Chicago, Indianapolis & Louisville, with three more additional sections of work under the second Commission order, with a total mileage of only 778 track miles and 226 locomotives to equip.[38]

Representing more than two decades of work, and personal expenditures that probably reached close to a million dollars when he finally closed down the Sprague Safety Control & Signaling Co. in 1933, Frank Sprague's venture in railway signaling had certainly not been a profitable one for him. As one of the early advocates of automatic train control, and a tireless promoter and advocate of it, Sprague could certainly share some of the credit as one of those who made it possible. And his early development of well-designed and -built equipment for its use undoubtedly had benefits that went far beyond the equipment that Sprague himself built.

Reflecting upon his experience with automatic train control in a 1934 publication of the American Institute of Electrical Engineers, Sprague summarized this part of his life in a rather apt understatement:

> The period of activity in this field, extending from about 1911 to 1933, was one of great trial and strain, and considering the commercial results, might have been more advantageously spent in some other direction.[39]

On October 13, 2008, President George W. Bush signed into law the Rail Safety Improvement Act of 2008 that required, among other things, the completion of Positive Train Control (PTC) on all Class I and intercity passenger and commuter railroads by the end of 2015. The legislation passed swiftly after a train collision in Chatsworth, California, on September 13, 2008, resulted in 26 fatalities, caused by the failure of a locomotive engineer to follow signal indications. PTC includes a modern version of the automatic train control system originally developed by Frank Sprague almost a hundred years ago.[40]

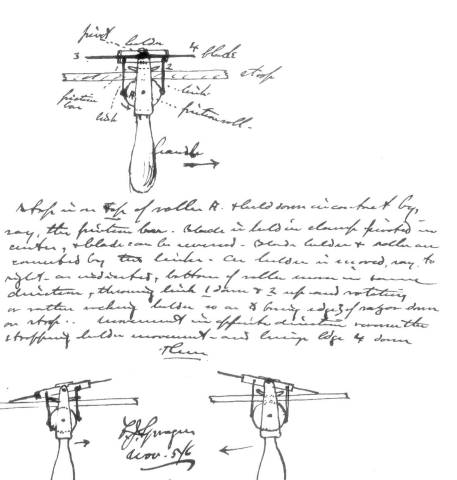

Sprague's design for stropping a razor blade. This allows for a back-and-forth motion to strop the blade. *Frank Sprague Papers, Manuscripts and Archives Division, The New York Public Library, Astor, Lenox and Tilden Foundations.*

10

A DIVERSE INVENTOR

As an inventor, Frank Sprague presents us with a complex character of sometimes seemingly contradictory traits. On the one hand, he provides a textbook example of the "inventor's shop" model of focused, directed research on a specific set of design problems—working with his colleagues and employees methodically testing and revising designs in a disciplined shop environment. On the other hand, he also displayed characteristics more in accordance with the "lone inventor" stereotype—jotting down ideas and plans as they occurred to him, on nearly any design problem that presented itself during his daily business. Throughout his career, Sprague relied on both spontaneous creativity in recognizing and meeting design challenges, and disciplined, methodical work in refining his ideas. He combined both of these traits with an indomitable sense of purpose and tireless zeal for pursuing and promoting his ideas, as well as asserting his priority to specific inventions or design elements, particularly when he believed himself to be in the right. He must at times have seemed to his "opponents," and probably to some of his colleagues as well, as something of a gadfly. He did not accept failure easily, and at times persevered against the odds to his cost. We will return to this aspect of Sprague the inventor below.

It is clear that even at the start of his career as an inventor, while still a midshipman in the U.S. Navy, Sprague had already established the combination of spontaneity and discipline as his modus operandi—although, of course, his way of working would continue to evolve throughout his career. He kept careful notes of his ideas and tests in diary-like notebooks, but at the same time jotted down ideas on miscellaneous scraps of paper, hundreds of them, including the opposite side of an order of court martial for an unfortunate colleague.[1] This was a habit that Sprague continued to the end of his career, but the air of chaos that it might imply is belied by the fact that each note and drawing is carefully signed and dated, and that Sprague preserved all of them. In an undated summary (probably completed sometime in 1885 or after), Sprague lists 57 inventions that he developed while still a midshipman in the navy between 1877 and 1880.[2] The inventions include such diverse items as a pocket phonograph, combination refracting telescope, electric control of torpedo, duplex telephone, octoplex telegraph, and a steering propeller, and have datelines such as "Yokohama," "China Sea," and "Singapore." Although an interesting list, none appear to have been patented.

Sprague filed his first successful patent application (no. 304,145) for a "Dynamo-Electro Machine" in 1881 while still with the navy in Newport, Rhode Island. In another undated summary of his early career (also probably completed in 1885 or later) he points out that the device could be used either as a motor or a generator.[3] The patent was issued August 26, 1884, after a later patent application (no. 295,454) for an "Electro-Dynamic Motor," applied for in 1883 after he had left the navy.

THE SPRAGUE "SYSTEM"

Sprague was already a fledgling inventor when he left the navy in 1883 and began working for Thomas Edison as a technical aide. Although Sprague was soon to run into problems with Edison that would occasionally trouble both men for the rest of their lives, Sprague undoubtedly benefited from the association. By the time Sprague joined him, Edison had already established the "inventor's shop" as a model for the development and production of new inventions. It is perhaps ironic that although Sprague personally had difficulties accepting this arrangement in the subordinate role, he too successfully implemented the shop model in his later work. Perhaps the difference lay in Sprague's management style.

The journals and shop logs kept by Sprague and his colleagues show that Sprague ran a meticulous operation. Six-day work weeks seem to have been the norm, and the work appears to have consisted of periods of development, production, and testing. Sprague and his key employees (such as Harold Cockerline) all kept experimental journals in which the results of any laboratory or field tests were recorded for any given day. These journals show a pattern of intense testing of new prototypes, revision of prototypes, and further testing, followed by sometimes months of relatively little experimental activity. Often, data from new tests were plotted alongside early tests to facilitate the comparison and evaluation of the results. Sometimes the journals show that experimental work was interrupted or delayed by the necessity of preparing documentation for patent interference cases.[4]

The shop logs were even more important for Sprague's work. In these, he recorded both the original designs for his inventions, as well as each subsequent modification thereof. Each new revision to the design would be pasted onto the appropriate page of the log as a fold-out leaf and signed and dated by those present when the amendment was made. This made it possible to review the course of development of a particular design element, as well as providing documentation that could be used in potential patent infringement suits, which were numerous. Many of the exhibits that Sprague and his patent attorneys used in litigation were taken directly from the shop logs, and parts of several logs still have court seals on them.[5]

Sprague also dictated exhaustive notes on his ideas for designs to his personal secretaries. These notes were sometimes revised, in their entirety, on an almost daily basis. In these dictations, which were duly signed and dated by Sprague and witnesses, he outlined in minute detail various ideas, potential designs, and modifications thereof. An example is a series of dictations he made to B. McLaughlin between October 18 and 26, 1911, for a reversible razor strop (which he ultimately did not patent).[6] Other examples include "Emergency Generator for an Aeroplane Radio (1928)," "A Device for Saving Power in an Electric Car Generator (1911)," "Some Notes on a Helicopter Type of Aeroplane (1916)," and "Notes on Canal Haulage (1903)."[7]

The most interesting aspect of the "Sprague system," however, is his previously mentioned, seemingly spontaneous sketches for various design ideas. These were drawn on everything from random scraps of paper (like the above-mentioned order of court martial), to Sprague company and hotel stationery, to used daily comparative price registers.[8] From the diversity of his writing materials one might imagine that he was a precocious member of the conservation lobby, but it is more likely that he simply grabbed whatever material was at hand when an idea occurred to him. The breadth of ideas he expressed in his sketches is even more diverse than the various materials he drew them on. They of course include the complete range of electrical inventions for which he is best known, but also items such as pens, floor sanders, razor strops, ordnance, and overhead tractors for canal barges.[9] Often, the sketches are rough,

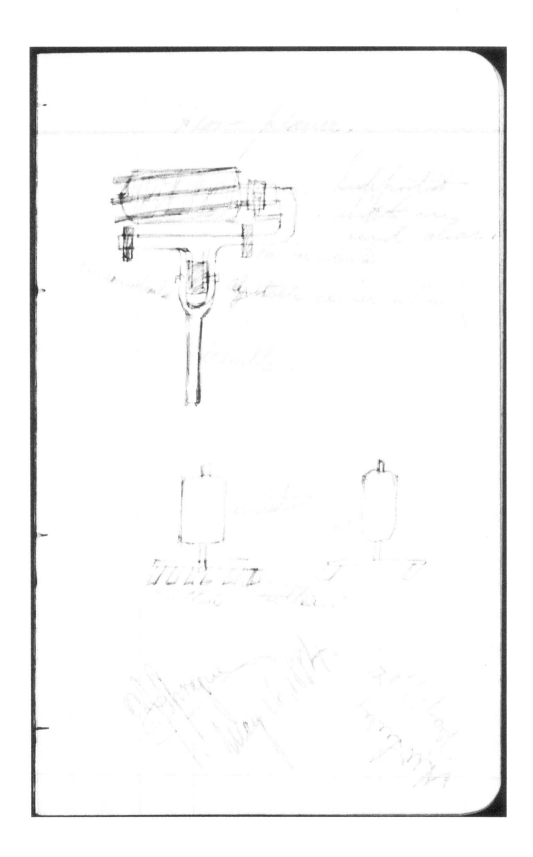

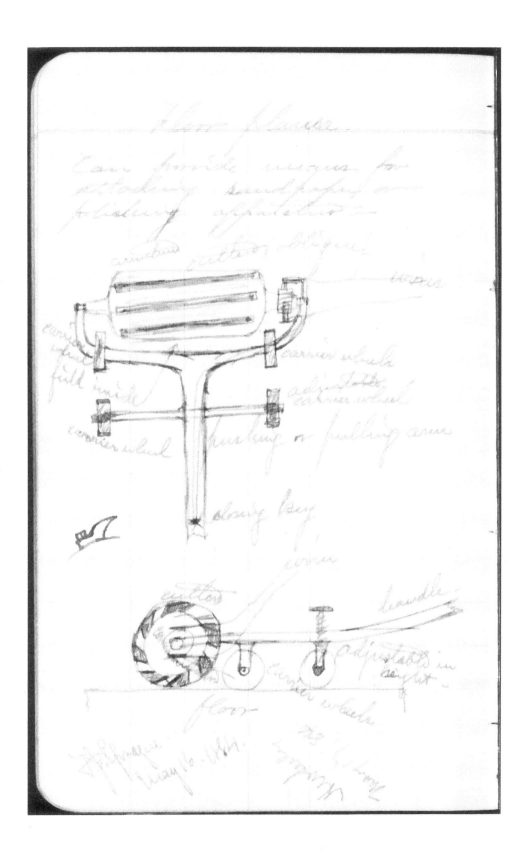

freehand renderings, and many are stained by stray drips and smears of ink. However, as often as not, they are so well drawn as to seem drafted, regardless of the material they are written on. However spontaneous these drawings seem in retrospect, each (as previously noted) was carefully signed, dated, and frequently stamped, for authentication.

Most of Sprague's sketches do not appear to have progressed much beyond the initial phase, although some, such as his ideas on canal haulage and the razor strop, appear to have received concerted attention over an extended period of development. Others, of course, were eventually patented. Overall, the sketches might give the impression of chaotic, even manic, creative energy, but the care that Sprague took to document the process suggests otherwise. It is much more likely that Sprague, like other productive, creative individuals in many other walks of life, produced an endless stream of new ideas. It seems that he recorded these ideas, more or less as they occurred to him, for future reference. Many, probably most, were abandoned or simply not pursued, others developed somewhat, and only a few pursued to patent and/or commercial application. This view of Sprague's creativity is much more in keeping with the other aspects of the "Sprague method."

COLLEAGUES

As Sprague's fortunes and commercial interests waxed and waned throughout his career he employed a sometimes considerable work force in his shops and factories. However, in addition to employees, he also employed individuals whose role could be better described as colleague. These relationships extended over years, even decades, and multiple projects, and were characterized by mutual respect, loyalty, and even affection. A number of Sprague's colleagues went on to launch respectable independent careers, but maintained close ties with Sprague up to his death. These colleagues included Oscar T. Crosby, who resigned from his army career to head Sprague's work at the New York plant for Richmond and elsewhere, while S. Dana Greene left the navy to head Sprague's field work in Richmond. Pat O'Shaughnessy was a capable

(*previous pages*) Two sketches copied from a May 1884 notebook, of a Sprague design for a new floor planer machine.   *Frank Sprague Papers, Manuscripts and Archives Division, The New York Public Library, Astor, Lenox and Tilden Foundations.*

(*right*) In this May 21, 1884, sketch, Sprague was working out the circuiting for an electric motor.   *Frank Sprague Papers, Manuscripts and Archives Division, The New York Public Library, Astor, Lenox and Tilden Foundations.*

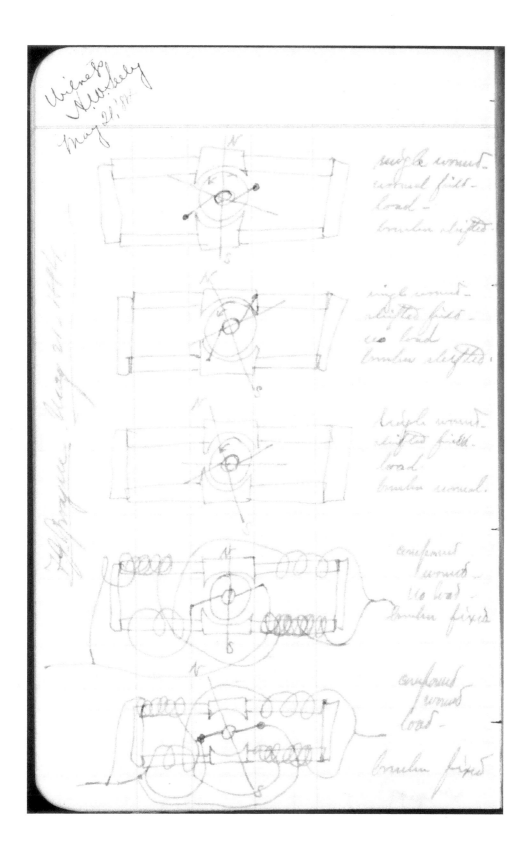

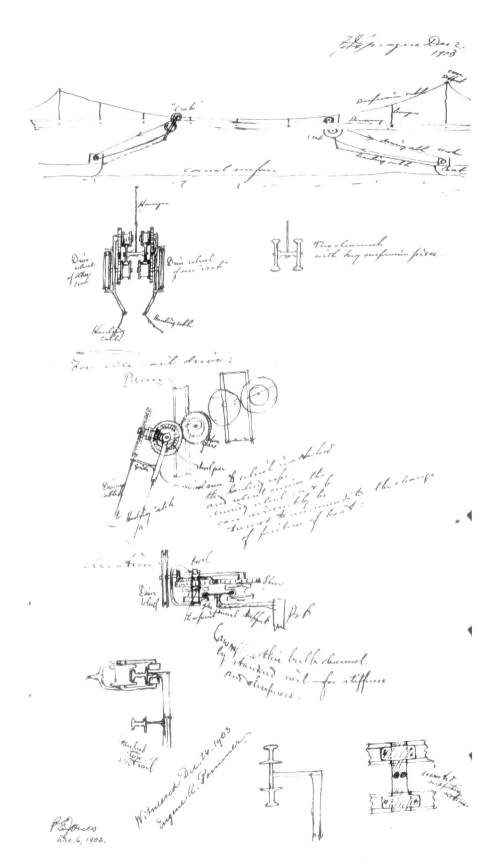

and resourceful mechanical man who worked with Sprague for some 20 years. Other colleagues included H. B. Cockerline, M. Knab, W. L. Hauck, and F. M. Shannon, as well as Sprague's son Desmond.

Desmond was one of his closest colleagues during the latter part of Frank's career. Desmond worked for his father in a variety of capacities on all of Sprague's most important endeavors over the last three decades of Frank's life. The relationship appears to have been a very close one, and their correspondence is a cozy mixture of family news and hard-nosed business. It also shows that their working relationship was based on more than simple filial piety—on a number of occasions, Desmond chides his father over his stubbornness or refusal to follow Desmond's suggestions, while Frank too found occasion to rebuke his son.

Desmond's role varied with the occasion and the needs of the current project, and included shop and lab assistant, administrative assistant, branch manager, and field supervisor. For instance, Desmond lent considerable support to Frank's efforts on the Naval Advisory Board, at times working directly with Frank, at other times managing Frank's field experiments while Frank was otherwise occupied, and at times pursuing his own independent (but related) work in naval ordnance, work that ultimately led to several of Desmond's individual patents.

Another of Sprague's important colleagues during the latter part of his career was Harold B. Cockerline. Cockerline worked with Sprague from about the 1910s to the mid-1920s, also in a variety of capacities. Most of his work appears to have been focused on experimental work in the shops, and much of that was focused on the development of the train control patents. That Sprague valued Cockerline's contributions to his efforts is made clear in a letter of recommendation Sprague wrote for Cockerline, in which he describes Cockerline in the most laudatory terms and gives him the highest possible recommendation.[10] That the respect was mutual is evidenced by the fact that Cockerline continued working with Sprague for almost another decade.

RICHMOND

Sprague's electrification of the Richmond Union Passenger Railway was his first major success, conclusively demonstrating the efficacy of electric power in urban transit, launching a boom in the construction of electric street railways,

Frank Sprague had a keen and long sustained interest in developing canal boats. This December 2, 1903, drawing shows ideas for powering cable boats.   *Frank Sprague Papers, Manuscripts and Archives Division, The New York Public Library, Astor, Lenox and Tilden Foundations.*

and establishing himself as a leading figure in electric traction. It also was a model that he sought to repeat in some of his later ventures.

At the time of the Richmond electrification, there already were several electric street railways in operation and several other projects under development. In general, they shared two common problems: delivering motive power to the wheels of the cars and drawing electric power from an external source. The problem with motive power is that the wheels of the cars moved independently of the chassis. Most designers mounted the motor on the chassis and used belts, chain drives, and a variety of other mechanisms to indirectly transfer motive power to the wheels. Sprague had already solved this problem before he started the Richmond project with his so-called "wheel-barrow," or nose-suspended motor mount, which he had originally developed for the electrification of the Manhattan Railway (which ultimately was not undertaken at that time). This method of mounting an electric motor suspended one end of the motor on a spring mounting from the truck frame, while the other was supported directly from the axle, allowing for the continuous delivery of motive power to the car's wheels despite irregularities in the car's movement.

This method of motor mounting was not originally from Sprague. The idea had been claimed as early as 1871 for a steam-powered road wagon, and in 1882 Dr. Joseph Finney described a similar arrangement in a streetcar in an application for a patent, which was denied by the Commissioner of Patents because "it is old to support the motor of a car in the manner selected by the applicant." When Sprague applied for a patent in 1885 it was not for the motor mounting itself, but rather for a "new and useful Improvement in Electric Railway Motors" (Electric Railway System, 324,892 Application date May 25, 1885, Issued August 25, 1885), which was successful.[11]

In what was to become a hallmark of the "Sprague system," he undertook the contract without having fully worked out the other problem, that of how the cars were to draw power. He also undertook the contract at his own financial risk.

By the time Sprague began to work on the problem, his "rivals" had already experimented with a wide array of solutions, ranging from third rails, underground conduits, and two-wire overhead systems. None of these were terribly reliable, and some presented distinct safety hazards. Sprague too experimented with a variety of systems, and ultimately settled on a single-wire, under running, wheeled trolley pole. Unfortunately for Sprague, Charles Van Depoele also had been experimenting with such a system, and the two ultimately ended up in court fighting a patent infringement suit.

According to Sprague, he had already solved the problem in 1882, although he didn't file for a patent for several years. However, the drawings for his 1885 patent application (Electric Railway System, 408,544, applied for February 27, 1885, granted August 6, 1889) show that he still had some way to go (and the system depicted in the patent is not the one that ultimately was used in Richmond). A sketch of his, dated November 20, 1888, shows that he was still experimenting with a variety of formats.[12] The suit was decided in Van Depoele's favor, and Sprague did not take the decision well. In one of his undated résumés, Sprague described the situation:

> 1882: . . . While in London devised the *under-running trolley* system, the basis of all modern trolleys, this comprising an overhead central rail following the center line of all tracks and switches contact to be made by universal self-adjusting upward-pressing contact carried over the center of the truck with return circuit through the rails. This was the origin of the modern trolley system, but application for patent therefore was not filed until three years later, resulting in an interference with the late Chas. W. Van Depoele. In the attendant proceedings, although Sprague was a Naval Officer on duty at the time of the invention, and shortly afterwards on an American Man-of-War and hence on American soil, he was placed by the Patent Office in the same class as a foreigner and his testimony as to conception limited to his return to the U.S. on May 30, 1883. This unjust decision not being contested by his attorneys, patent was permitted to go to Van Depoele, but with free license for Sprague.[13]

In spite of the ultimate setback with the patent on the trolley pole, Sprague did finally come up with a version of a current-collecting wheel mounted at the end of the pole that solved many of the problems with powering the cars. Sprague had a number of other problems to contend with, including the gearing and brushes of his motors, all of which he successfully overcame by dint of hard work, trial and error, and constant revision. He also had the help of several colleagues, in particular S. Dana Green, Oscar T. Crosby, and Pat O'Shaughnessy (the latter with whom Sprague shared authorship of patent 414,172, granted in 1889).

When the Richmond Union Passenger Railway began operating in 1888, it was one of only a handful of electric street railways in the country. However, Sprague's success in mastering two of the abiding design problems demonstrated that electric traction was a viable option for urban transportation. Within just a few years, the number of electric street railways in the United States had increased tenfold, and Sprague's company built many of them.

F. J. SPRAGUE.
ELECTRIC RAILWAY SYSTEM.

No. 408,544.   Patented Aug. 6, 1889.

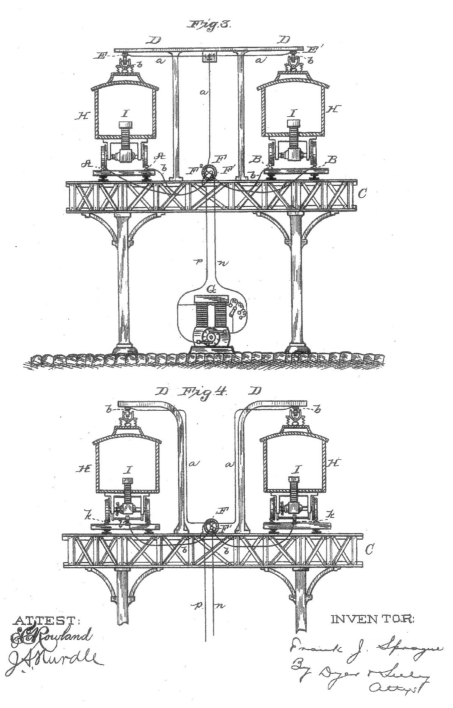

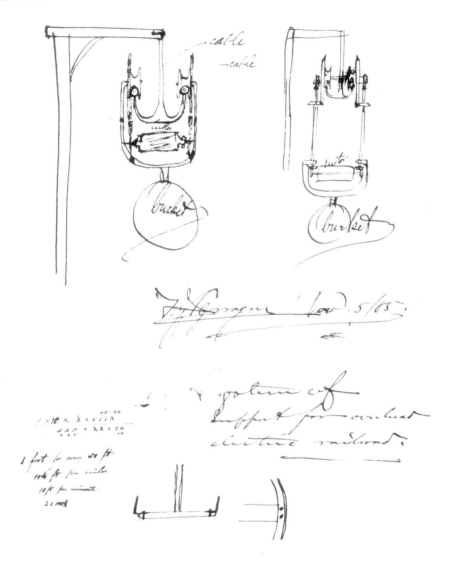

On November 5, 1885, Frank Sprague was still trying to find the best arrangement for an overhead trolley wire. (*left*) Sprague's earliest design was the one he developed from his visits to the London subway in 1882, which was applied on February 27, 1885, and patented in 1889. This used an under-contact current collection from an overhead rail in the subway.   *U.S. Patent and Trademark Office.* (*above*) On November 5, 1885, Sprague developed these sketches for a two-wire overhead system on a piece of stationery.   *Frank Sprague Papers, Manuscripts and Archives Division, The New York Public Library, Astor, Lenox and Tilden Foundations.* (*following page*) On November 30, 1885, he developed sketches for a single wire arrangement with V-shaped contacts at the upper end of the pole.   *Frank Sprague Papers, Manuscripts and Archives Division, The New York Public Library, Astor, Lenox and Tilden Foundations.*

Sprague's self confidence had paid off, and although he ultimately sold his interest in the company and turned his attentions to new projects, he would return to the problems of electric traction in less than a decade.

THE MULTIPLE UNIT SYSTEM

Although Sprague took pride in many of his accomplishments, without a doubt, the multiple unit system was his most significant invention, both in terms of its practical ramifications and Sprague's individual fortune. It also provides us with a good example of the Sprague system successfully employed to its fullest advantage. In particular, we see how his thorough, methodical approach, his dogged perseverance, and his supreme self-confidence paid off.

Sprague's own view on the matter is informative. In an undated resume (prepared near the end of 1904 or later) regarding the problems faced by the

New York subway and other electric railways, he cites three key events that made electric railways in general, and underground railways in particular, possible.[14] He begins by pointing out that without electricity, subways and underground railways would be impractical, and follows this with the observation that "the testimony of history is practically unanimous that the installation of the Richmond Road by Sprague in 1887–88 was the most potent single influence beginning the modern development of electric traction." His second point is that the multiple unit system made possible "the high schedule speeds and ultimate maximum capacity of the subway and other transit roads" and that the system "now universally adopted, was invented by Sprague, and inaugurated by him on the South Side Elevated Railroad of Chicago in 1897–8." Sprague's third and final point is that without his own "constant effort, public teaching, and willingness to demonstrate the possibilities of electric transmission and tractions" there would have been delay in the implementation of electric traction, and that he had always been ready "to back his opinion financially as well as professionally." These three points highlight the different aspects of Sprague's creative process that made the multiple unit patent so successful for him.

Sprague traces his involvement in electric traction to the summer of 1881 and his first research on and patent application for a dynamo-electric machine. In the years that follow (the resume runs through August of 1904) he enumerates an impressive list of inventions and innovations, but these make up only a portion of the events Sprague himself considered to be the most significant in the development of electric traction.

Sprague appears to have been following the same strategy that led to his success in Richmond in pursuing the implementation of the multiple unit system. He repeatedly made offers to various municipalities "at his own expense" to install a test system to demonstrate its efficacy. These offers were often at first rejected, much to Sprague's frustration, and one can almost feel the apparent sense of satisfaction when Sprague was able to note that the multiple unit system was eventually adopted by a recalcitrant municipality.

His comments reveal that he was not simply peddling his system, but the system that he clearly believed to be the superior one for urban traction. His account of his many articles, interviews, public presentations, offers to equip roads with test equipment at his own expense, and presentations to municipal and other authorities are peppered with comments that reveal both an unshakable sense of self-confidence and a conviction that he had the superior solution:

Took contracts for the St. Joseph and the Richmond Union Passenger Railways, the completion of the latter marking the beginning of the modern development.[15]

Spurred on by manifest obstructions, he made, in the *Evening Post,* an unusual effort designed to stop all extensions of the Manhattan Elevated, and to advance the construction of an underground railroad and its equipment with electricity.[16]

In an interview in *N.Y. Evening Post* . . . advocated underground road equipped on a new plan, the "multiple unit" . . . pointing out the advantages of increase schedules, reduced strains and higher economies. This was reduced to practice within two years, and the system has now been universally adopted.[17]

In *New York Times,* criticized pessimistic attitude assumed by a leading manufacturer in relation to the demand for electrical operation of trains in the New York Central Railroad tunnel, and upheld multiple unit operation.[18]

As result of multiple unit missionary work, general specifications for the New York Subway motors and schedules were issued, the latter being practically those specified by Sprague four years earlier.[19]

Richmond made possible the individual car, but the South Side of Chicago the *electric train,* and although the first public trial of the multiple unit system, vital not only for increasing schedules but to get the ultimate capacity oft of a given trackage, was made less than eight years ago, its universal adoption marks a new marks a new era in electric railway operation.[20]

A final aspect of the multiple unit system that sheds some light on Sprague the inventor is the multiple unit patent itself (no. 660,065, applied for April 30, 1898, granted October 16, 1900). The document is comprehensive, with 263 claims (unique, protected elements of the patent) covering every aspect of the system. His patent was robust enough to withstand all of the numerous patent interference suits with which he had to contend, and ultimately forced all of his competitors to deal with him. Sprague licensed the patent in the United States and Europe, assuring his fortune. His success with the multiple unit patent also became a model for some of his future endeavors, which, unfortunately for Sprague, would not be so successful.

AUTOMATIC TRAIN CONTROL

Automatic train control was Sprague's next major railroad project following the success of the multiple unit patent, and although he ultimately was granted a patent for automatic train control, we might say that in this case, the Sprague system backfired. Outside events including political and regulatory machinations, the existence of viable and unique competition, and the start of the Depression all contributed to a less than favorable climate for success, but a key element in Sprague's failure with automatic train control was a part of the Sprague system itself: his sense of determination to persevere against the odds with the conviction that he had the superior system.

Sprague's success in Richmond had shown him that audacious risks could pay off, and he repeated this with the multiple unit system on Chicago's South Side. His success with the multiple unit system appears to have taught him that a truly comprehensive patent would force his competitors to deal with him, no matter how large or financially powerful they were, and this is where the Sprague system failed him. Instead of a relatively rapid approval of his patent such as he had had with the multiple unit system, Sprague found himself engaged in a protracted series of patent infringement suits and fights with the patent office, many of which he lost. By the time these conflicts were resolved (or Sprague had capitulated) and the patent issued, Sprague was left with a less comprehensive patent than he had desired and was confronted by an economic and regulatory environment that held little demand for a new train control system (one of several that were available at the time).

Automatic train control was, in a sense, the culmination of Sprague's life-long work in traction and railroading. As he envisioned it, the system would (automatically, of course) override violations of a train's operating parameters, whether caused by human error, mechanical failure, or even, as we may infer from a newspaper clipping pasted into one of his shop books, the untimely demise of the operator (it seems that Sprague meticulously clipped every article about horrific railroad accidents that presumably could have been avoided had his system been in use).[21] He was often at pains to point out that, despite the name, the system would not obviate the need for an engineer, but only activate when an engineer failed to follow the operating parameters.[22] A number of Sprague's earlier patents anticipate various elements of the system, but as with the multiple unit system, he wanted a comprehensive patent that would consolidate all of the disparate elements into a single system.

Although Sprague had obviously done considerable work on automatic train control before he filed for his patent, the real drama of the story begins with his filing on December 31, 1914 (the date of formal acceptance of his application by the Patent Office). Almost immediately there followed a series of exchanges with the Patent Office, often verging on confrontational, over various aspects of the application. The Patent Office raised a considerable

A scene from Frank Sprague's workshop, where a GE brake system had been set up to test elements of Sprague's automatic train control system. He would typically test the various elements of the system in his shop before deploying them for further field tests. The lettering was added later when he used the photo in one of his patent interference cases.   *Frank Sprague Papers, Manuscripts and Archives Division, The New York Public Library, Astor, Lenox and Tilden Foundations.*

number of challenges to the application, ranging from technical aspects of the application, to how figures should be drawn and numbered, to grammatical corrections.[23] Sprague balked at nearly all of the suggestions, but did make a number of amendments to the application.[24] The Patent Office replied that the amendments were not sufficient, and that unless their concerns were addressed by the following year, the application would be considered to be abandoned.[25] Correspondence between Sprague and the Patent Office continued in this vein for much of the next 15 years, although the intensity of the exchanges waxed and waned.

To the modern reader, many of the Patent Office's objections seem astonishingly petty, constituting grammatical and editorial corrections that would have had no substantive impact on the patent itself. Equally astonishing is Sprague's steadfast refusal to concede them. The Patent Office also raised more substantive challenges, particularly over specific claims in the application.

Each claim, in a sense, staked out more exclusive territory for Sprague, and as such, these were substantive, and he defended his claims with even greater vigor than his grammar. In the end, he would concede most of these points to the Patent Office.

In addition to the Patent Office, Sprague also had to battle other inventors in dozens of patent infringement suits. In some of these, he was contesting with earlier patentees, striving to establish that his inventions constituted novel and substantial improvements on the state of the art. In others, he fought against subsequent patent applications, arguing that they impinged on his own work.[26] Some of these suits he lost, and with them, lost important claims staking out his exclusivity to various aspects of the automatic train control system. Others he won, gaining him, at least temporarily, a hold on other important claims.

Throughout this period, Sprague was also hard at work on other projects, including his elevator business, his safety control and signal business, and the Naval Advisory Board. However, he continued to spend considerable effort on the continued development of the automatic train control system. His notes, shop records, and experimental journals show that he was continuously revising, testing, and adapting various elements of the system. Many of the tests were run on New York Central locomotives, and Sprague and his colleagues kept detailed notes on their maintenance and performance.[27] The records also mention numerous tests of the system performed for various visiting railroad officials, and the installation of test systems on several roads, but with seemingly little result.

Sprague continued his battles through 1929, showing the same dogged determination and conviction in his own righteadedness that he had exhibited during his struggles to establish the multiple unit system as the sine qua non of electric traction, but with far less to show for his efforts. In May of 1929 he received a letter from the Patent Office that baldly stated:

> This application has now been pending for (14½) fourteen and one-half years, and has long become special, under commissioner's order 2210 of June 28, 1915. It is submitted that, in the multitude of claims found allowable in this application, applicant has secured ample protection for the invention disclosed in this application, and closure of the ex parte prosecution of this application before the Primary Examiner is now well in order. It is herewith requested that applicant, in replying to the present action, enter such response as will terminate the ex parte prosecution of this application before the primary examiner.[28]

Now near the end of his life, Sprague decided to capitulate. He conceded all of the remaining challenges raised by the Patent Office, including several that he had been vigorously fighting since 1915, some of them merely grammatical, and the patent was issued (no. 1,780,148, filed December 31, 1914, granted October 28, 1930). The patent allowed a total of 155 claims, not as many as the multiple unit patent, but still a considerable number. Sprague probably viewed this as more of a defeat than a victory.

The circumstances in which Sprague found himself on winning his patent were far different than those under which he began his efforts. The fortune he had won with the multiple unit patent was depleted, he faced competitors with deeper pockets who had developed alternative systems of automatic train control, the regulatory pressure for railroads to adopt automatic train control had abated, and the Depression had begun. There simply were very few takers for Sprague's system. Although Sprague's determination and self-confidence clearly helped him win the multiple unit patent, they seem to have cost him his ultimate victory in the case of automatic train control. His nearly 16-year fight had delayed his entry into the market until the market no longer existed.

Frank Sprague's 75th birthday celebration was an extraordinary event, held on July 25, 1932, at the Engineering Societies Auditorium in New York. Frank and Harriet looked over the six bound volumes of letters and photographs presented to him at the occasion. *Courtesy of John L. Sprague.*

11

AN INVENTOR AND
ENGINEER TO THE END

In July 1927 Frank Sprague moved into the 70th year of his life, and one might have expected him to begin easing up on the level of his work, or to have begun to enjoy the pleasures of a life of semiretirement. But this, of course, would not have been Frank Sprague. From the time of his youth onward he had always held these strong interests in an extraordinary range of diverse topics, and he would hold them throughout his life.

Sprague, working with his eldest son, Desmond, would continue his long-running work on his Sprague Safety Control & Signaling Corp. until well into the 1930s. He was in his 69th year when he began work on his innovative dual car elevator design in 1926. And before the end of the decade he would begin his work on his patented Universal Electric Sign System which would use massed electric lamps to display a great variety of signs in either still or moving arrangements, and which could move in different arrangements and at different speeds. At least one example of the Sprague sign technology used was a large moving sign that he designed as part of the Time-Fortune exhibit at the 1933–1934 Chicago Century of Progress Exhibition.

Two other ideas, not yet fully developed, were jointly pursued by Sprague with L. C. Sprague (no relation), an old friend who was the general manager of the Uintah Railway of Mack, Colorado. In 1925 the two had begun work on ideas for an advanced version of steam locomotive boosters on tenders. The two men developed two basic alternatives for the boosters that were described in a lengthy outline in a December 2, 1924, letter to L. C. Sprague.

One version would operate with the tender booster unit completely separate from the power of the locomotive boiler. A separate high speed Diesel engine, a generator, and traction motors would all be carried on the locomotive tender. This variation of the proposed booster was discarded from consideration because of the still developing nature of the Diesel engine. Instead, they planned to use a steam booster unit driving a generator and traction motors. This arrangement would draw excess steam from the locomotive's main boiler, powering an electrical generator and traction motors installed on the tender. The supplemental booster would automatically use excess steam available when an engine was either accelerating from a start, or when it was nearly stalled on a heavy grade. The booster would otherwise operate only when extra power was needed and sufficient excess steam capacity was available.[1]

By 1931 Frank Sprague had begun to develop a design for a rail detector car, an improved version of the gyroscopic rail detector car originally developed by Elmer Sperry, and Sprague and L. C. Sprague were looking for others that might be interested in the new project.[2]

Neither of these ideas was ever fully developed.

While Frank Sprague continued to entertain a wide range of technical topics that encouraged his involvement, the worsening Great Depression that followed the late 1929 stock market crash was hardly a good time for the development of new ideas or their financing. Sprague had already expended a substantial sum for development of the Sprague Safety Control & Signaling Corp., which had been under way since 1907 with very little remuneration ever received from the work. And his ambitious ideas for electric sign development were also requiring significant financing.

Sprague had earned a substantial amount for his sale of the multiple unit control design to GE in 1902. But none of his later projects had paid off nearly as well, or many of them not at all, and by the 1930s his finances were nearly diminished. Some idea of the condition of his finances was suggested in a Christmas letter to Mary Das, Frank Sprague's first wife, in 1933.

> My own developments in the Sign business—in which there has been a very heavy expenditure, considerably over $150,000, have been important from a technical standpoint, and if I can carry on there ought some day to be some commensurate return. But the present business conditions are most discouraging for any new investment or risk. I am however optimistic, and I hope to fool the financial doctors in time to leave some tangible values.[3]

Beginning their work some 6 months in advance of Sprague's 75th birthday, an informal committee of some of his oldest associates and admirers began to organize a remarkable reception that would honor him as one of the most distinguished members of the engineering profession and to proclaim his achievements to the world. This informal committee included Frank H. Shepard of the Westinghouse Co., William B. Potter of the General Electric Co., Frank Hedley, President of the Interborough Rapid Transit Co., and Guy A. Richardson, President of the American Electric Railway Association. A larger Frank J. Sprague Anniversary Committee included the presidents of several engineering societies, the presidents of universities which had given Sprague honorary degrees, a number of Sprague's early friends and associates, and many men prominent in electrical science and industry. Chairman of the Anniversary Committee was Gano Dunn, President of the J. G. White Engineering Corp., who presided over the meeting.

The committee put together a remarkable group of letters and photographs from Sprague's family, friends, and associates going back all the way to his years at the Naval Academy, with nearly 500 of them received. Their responses came from all over the United States and abroad.

His navy associates were represented in large numbers, including his classmates James H. Glennon and Harry McL. P. Huse, who had both retired as admirals, and Mortimer E. Cooley, who had become Dean of the College of Engineering at the University of Michigan. Among those who had worked with Sprague on his many projects were Oscar T. Crosby, who had been one of Sprague's key men in the successful project in Richmond; Charles R. Pratt, who had worked with him on the development of electric elevators; and Dr. Cary T. Hutchinson, who together with Sprague and Dr. Louis Duncan had developed the pioneer Sprague electric locomotive of 1895.

Among the many engineering consultants who celebrated Sprague's 75th anniversary were noted bridge engineers Gustav Lindenthal and Ralph Modjeski. New York Central vice president and chief engineer William J. Wilgus, and consultants Bion J. Arnold and George Gibbs, who had served with Sprague from 1902 to 1906 on the railroad's Electric Traction Commission for the New York Central's New York electrification. Some time later Wilgus, who had been unable to attend the 75th birthday celebration, remembered his association with Sprague with particular affection.

To few men have such tributes come in their life-time. They must go far to offset the trials which to most of us now weigh so heavily. I do very heartily congratulate you on the recognition of your genius which has come to you while still this side of the great divide.

To me, in this connection, ever lives the memory of your first visit to me in 1899 as an advocate of the electrification of the New York Central; and of your succeeding association with me in the putting over of that remarkable change which has so amply borne out our prophesy.[4]

Heads of major companies that honored Sprague's long and distinguished career included General Electric President Gerard Swope, Westinghouse Air Brake Co. Chairman H. H. Westinghouse, Otis Elevator President Jesse H. Van Alstyne, and Baldwin Locomotive Works Chairman Samuel M. Vauclain. Overseas companies, railroads, and transit companies represented include Chairman Carl F. Von Siemens of the Siemens Companies, as well as several of his directors, the Right Honorable Lord Ashfield, chairman of the London Underground, and electrical engineers from companies in Belgium, Germany, and Switzerland. Railroads and public utilities companies included railroad presidents William W. Atterbury of the Pennsylvania Railroad and Ralph Budd of the Burlington; and transit executives Britton I. Budd of Chicago Rapid Transit, Thomas Conway, Jr., of the Philadelphia & Western, Edward Dana of the Boston Elevated, Samuel Insull, and Peter Witt.

Representing universities were such presidents as Dr. Nicholas Murray Butler of Columbia, Rear Admiral Ralph Earle of Worcester Polytechnic, Dr. Thomas S. Gates of University of Pennsylvania, and Dr. John Grier Hibben of Princeton. Cultural friends representing Frank and Harriet's long involvement in the arts included Ossip Gabrilowitsch, conductor of the Detroit Philharmonic Orchestra, and his wife, Clara, the daughter of Samuel Clemens; actor George Nash; author Booth Tarkington; artist August Franzen; composer and conductor Dr. Henry K. Harley; author and artist Oliver Herford; author Lyman Beecher Stowe; and poet Dr. Edwin Markham.

Public notables included President Herbert Hoover, New York Governor Franklin Delano Roosevelt, Great War leader General John J. Pershing, John J. Esch, the past chairman of the Interstate Commerce Commission, Secretary of the Navy Charles Francis Adams, and the former Secretary of the Navy Josephus Daniels.

The day before the anniversary, the *New York Herald Tribune* marked the occasion with a laudatory feature, "Father of Rapid Transit," by writer Robert D. Potter, about Frank Sprague's many accomplishments, published in the newspaper's Sunday magazine section.[5]

The 75th anniversary recognition for Frank Sprague began at 8 PM on the evening of his birthday—July 25, 1932—at the Engineering Societies Auditorium, 29 West 39th Street, New York, with almost 1,000 in attendance, an especially large gathering considering the depths of the Great Depression. Speeches introduced by the chairman included an address by Dr. John A. Finley, Associate Editor of the *New York Times,* on "An Engineer's Contribution to the World's Welfare," who was followed by Frank Hedley, President of the Interborough Rapid Transit Co., speaking on "An Engineers Contribution to Transportation." Rear Admiral S.S. Robison presented a tribute to Sprague, particularly emphasizing both his contributions to the navy and his extraordinary work as a great engineer. At the conclusion of Robison's remarks, ushers brought forth six handsomely bound volumes of the many letters and photographs sent to Sprague for the occasion.

After a response from Frank Sprague remembering his career and the people he had worked with, Admiral Robison gave an account of Sprague's great accomplishments to a National Broadcasting Company–Blue Network broadcast that was carried over 18 different radio stations from places as far away as Miami, Houston, or Bismarck, North Dakota, beginning with a reading of President Hoover's well wishes to Sprague:

My Dear Mr. Sprague:

I send you my cordial felicitations on your 75th birthday and all good wishes for the future.

Your contributions to the development of the electric motor, followed by the application of electricity to street railways and to elevators, links your name for all time with that distinguished group of inventors and engineers whose pioneer work made possible so many of our present utilities, comforts and conveniences.

It is fitting that the engineering and scientific world should show general recognition of your noteworthy services to the electric art, and it is with regret that I find I shall not be able to be present at your anniversary celebration.

Yours faithfully,
Herbert Hoover

Robison then concluded his remarks with a quotation from a leading technical magazine:

Inventors there have been whose talents lay so wholly in invention, and so little in life, that they lacked either the driving force to get their ideas before the public, or the ability to make friends or at least believers who would attend to such details for them. But once in a while there appears a

Frank Sprague and his family gathered together for the birthday celebration. Seated are Frank and Harriet and, to their right, their daughter Frances Althea. Standing to the far right is Althea's husband John Jerome Tucker, while standing immediately behind Frank are his eldest son Frank Desmond and his wife Ruth. Directly behind Harriet is their son Robert, while at the far left are son Julian and his wife, Helene. Missing is Robert's wife Florence.   *Courtesy of John L. Sprague.*

man whose grasp on business affairs is equal to his superlative inventive ability,—an inventor, a fighter, a pusher—one who is not content to produce what is new and better, but who will fight to the last gasp and the last penny to force an uninterested and reluctant world to adopt what he has to give. Such a man is Frank J. Sprague, who more than any other one man must be recognized as having brought electric transportation into being.

The formal proceedings ended and the remarkable gathering assembled in the engineering societies lobby, where old friends and colleagues renewed their years together with the Sprague's until midnight. It was an extraordinary event.[6]

Sprague continued his electrical work, but after his 75th birthday it was interrupted more and more frequently by ill health. Even before the birthday celebration, in June, Frank was laid up in bed on doctor's orders for a time. Over the next two years he had several attacks of grippe (influenza), and in March 1933 he suffered for some time with a severe attack of the flu. In a 1933 Christmas letter to Mary Das, Frank told of his severe weight loss and illness over the last year.

> During this period I had three attacks of pneumonia, the last about three months ago a double case, complicated with a threatened heart condition. The last was of such character that both my Doctor, one of the best specialists—and one of the most personally agreeable ones in the city, together with my nurse gave me a probability of only forty-eight hours before a permanent change of residence.[7]

Sprague recovered from the double pneumonia, but his ill health continued. Early in 1934 he regretted to his correspondents his erratic attendance at his office, and he had to decline several important invitations because of his ill health. But later in the year he was able to attend an electric railway meeting in Cleveland that must have given him great satisfaction. The occasion was the annual meeting of the American Transit Association, at which several new design streetcars would be operated.

The American street railway industry had grown into an enormous and profitable business in the years that had followed Frank Sprague's development of the first commercially successful electric streetcar in Richmond in 1888, replacing the horse and cable cars that had preceded them. By the early 1920s street railway systems had reached the peak of their expansion and the trolley car had become an indispensible and universal part of the urban American scene. By a peak year in 1923 there were more than 1,000 electric railway companies operating in the United States and its possessions, and well over 60,000 streetcars in operation on some 29,000 miles of street railway trackage. The streetcars transported almost 13.6 billion annual passengers and brought in annual operating revenues of more than a billion dollars.

But in the years that followed World War I, the street railways had begun to experience some major setbacks. The costs of labor and materials rose rapidly after the war, and the local authorities who had to approve the typically 3-cent or 5-cent streetcar fares were reluctant to increase fares, and the streetcars saw serious reductions in their earnings. The development of mass produced automobiles took thousands of transit riders away from the streetcars, and the

In honoring Sprague's many achievements in electric engineering and invention, this bronze portrait bust was presented to him at the January 1934 annual meeting of the American Institute of Electrical Engineers. Together with Sprague was sculptress Florence M. Darnault. *Courtesy of John L. Sprague.*

development of modern motor buses began to take still more streetcar riders away. By 1930 streetcar ridership had dropped by almost 25 percent.[8]

Many street railway operators saw that improved equipment would be a key element to providing better service. The heavy, lumbering old streetcars that had served the street railways so well were seen as slow and uncomfortable compared to the modern automobiles and buses now available. There were a number of efforts to design improved cars starting in the 1920s, but by far the most important of these was what was called the Electric Railway Presidents' Conference Committee organized in 1929. Headed by well known traction executive Dr. Thomas Conway, Jr., the committee hired research director Professor C. F. Hirshfeld to design a completely new, modern streetcar. By the summer of 1934 Hirshfeld had completed a Model "B" PCC (Presidents Conference Committee) car, a prototype of the streamlined new car that incorporated such features as an extremely lightweight welded high-tensile steel body, spring suspended motors, eddy current brakes, magnetic track brakes,

In one of the last photographs taken before his death, Frank Sprague enjoyed the garden at Harriet Ford Morgan's home at Milford, Connecticut.   *Frank Sprague Papers, Manuscripts and Archives Division, The New York Public Library, Astor, Lenox and Tilden Foundations.*

"floating" type control, and improved heating, lighting, and ventilation systems. Rubber-cushioned wheels and rubber insulation made the car extremely quiet in operation, and its acceleration and braking performance were substantially better than any previous streetcar design.[9]

The new car spent a month in tests on the Chicago street railway system, and was then brought to Cleveland together with several other new cars to be shown and operated at the transit convention during the September meeting. "Your presence is considered most fitting," wired his old friend Frank Shepard from Sprague, who made all the arrangements and escorted Sprague on the overnight train to Cleveland.[10]

A group of more than 200 prominent guests were invited for the demonstration on the morning of September 24. In addition to the demonstrations, guests were given a short ride on the new cars and taken to lunch at the Cleveland Hotel. Thomas Conway presided over the luncheon meeting, explaining the greatly improved performance, while Joseph B. Eastman, federal coordinator of transportation, expressed the opinion that the improved car had an important role in larger cities. Conway introduced Frank Sprague as the "Father of Electric Traction," who voiced the hope that continued progress would be made and expressed the opinion that the "will to live" had enabled the industry to come through very difficult times.[11]

The PCC car proved to be an extraordinary success, and it contributed to rejuvenation and the extension of a number of street railway systems. By the time North American PCC car production ended in 1951, more than 5,000 of the PCC cars had been built, and at least 15,000 PCC cars, or cars using PCC patents, had been built for overseas companies.[12]

For Frank Sprague, his journey to Cleveland would prove to be his last.

Early in October, Sprague came down with another case of pneumonia. But even in his illness he remained involved with his current project, the new electric lighting design, asking that a new model be brought in for his review. As his illness worsened, his doctor asked that a night nurse be brought in to help with his care. When Harriet went in to say goodnight to him, Frank remarked whimsically, "They've fired you."

On October 22, a delegation had arrived to notify Sprague that he had been unanimously selected by the four national engineering societies as the recipient of the high honor of the annual John Fritz gold medal, for "distinguished service as inventor and engineer through the application and control of electric power in transportation systems." That evening Harriet Sprague read to him the notice of award. This gratifying news was one of the last he

A prototype of the radical new design of the street car was completed by the Pullman-Standard Co. in 1934, and more than 5,000 of the streamlined cars were produced over the next 15 years. Frank Sprague traveled all the way to Cleveland to see the demonstration of the new car in September 1934, only a month before his death.   *Middleton Collection.*

would receive before losing consciousness. On October 25, 1934, at about 3 AM, Frank Sprague died in his New York home.

Funeral services were conducted at New York's West End Collegiate Church at West End Avenue and 77th Street on October 27, with the Rev. Dr. Henry Evertson Cobb of the Collegiate Dutch Reformed Church and the Rev. Dr. Edward M. Chapman of New London, Connecticut, a Sprague brother-in-law, conducting the service. Burial followed two days later in the Arlington National Cemetery in Washington, D.C.[13]

Harriet Sprague survived her husband by 35 years. In addition to having created an uncommonly congenial family for her spouse and children, Harriet was a notable rare book and first edition collector and she exhibited Walt Whitman collections. She was also an author on collections, and was a mem-

This undated photograph of Harriet Sprague was taken some time after Frank Sprague's death. She survived her husband by 35 years, and assured the preservation of Frank's work, authored her book *Frank J. Sprague and the Edison Myth,* and continued her work on rare books, first edition collections, and Walt Whitman collections.   *Courtesy of John L. Sprague.*

ber of a number of literary and historical organizations. After Frank's death, she had gathered an extremely large collection of his papers which she gave to the New York Public Library, the Engineering Societies Library in New York (since transferred to the Linda Hall Library in Kansas City, Missouri), and, later, to the Shore Line Trolley Museum in East Haven, Connecticut. In 1947 she authored the book, *Frank J. Sprague and the Edison Myth*,[14] which took up Frank Sprague's long complaints concerning the often overstated claims concerning the relative electric railway inventions of Sprague and Edison. Harriet died in Williamstown, Massachusetts, on October 1, 1969, at the age of 93. She was buried with Frank at the Arlington National Cemetery.

Sprague's oldest son, Desmond, worked as an engineer for his father until shortly before Frank Sprague died. With few new engineering opportunities at the height of the Great Depression, Desmond went on to photography and other work until his death in 1957. The Sprague's son Robert followed his father to the Naval Academy, where he graduated in 1920 and continued with the navy until 1928. The youngest son, Julian, attended Yale, and the two brothers joined in 1928 to establish the Sprague Specialties Company that manufactured radio capacitors. The new company was moved to North Adams, Massachusetts, in 1930, with a $25,000 loan from Frank Sprague. Later known as Sprague Electric Co., the firm grew into a major supplier of capacitors and electronic equipment, becoming one of the largest electronic component manufacturers in the world, reaching an annual volume of some $500 million by the mid-1980s. The Sprague family no longer managed the firm after it was sold to General Cable Corporation in 1976, and then to Penn Central Corporation, and Sprague Electric was sold or liquidated beginning in 1987. Portions of the company still operate as part of Vishay Intertechnology and Sanken Electric. Julian Sprague died in 1960, while Robert died in 1991. All three of the Frank Sprague sons are buried near each other in the Westlawn Cemetery in Williamstown, Massachusetts.

Frances Althea, the Sprague's only daughter, married architect John Jerome Tucker of Connecticut. She died in 1956 and is buried in Woodlawn Cemetery in the Bronx, in New York.

A modern version of Frank Sprague's durable electric street railway is now springing up as what are usually called "light rail lines" or simply "streetcars." While the décor of the new cars reflects modern streamlined design, such basics as street railway tracks or the 600 volt overhead power are unchanged. This is a train of the new Lynx light rail system at downtown 7th Street in Charlotte, North Carolina, that began operating in 2007.   *William D. Middleton.*

12
EPILOGUE

FRANK SPRAGUE AND THOMAS EDISON

Frank Sprague and Thomas Edison were contemporaries as electrical engineers and inventors in the exciting new world of electricity in the latter part of the nineteenth century. They sometimes worked in cooperation, and sometimes in competition with such great early electrical engineers as Nikola Tesla, Alexander Graham Bell, Elihu Thomson, George Westinghouse, or Charles Steinmetz. From the time the two men met in June 1878, just after Sprague's graduation from the U.S. Naval Academy, over the next half-century until Edison's death in 1931, the relationship between Sprague and Edison was one that varied widely. Working together and supporting each other sometimes, engaging in angry disagreements at others, they always acknowledged each other as great electrical engineers and inventors. They had much in common, both had sharp inquisitive minds, and both would prove to be extraordinarily inventive men.

The two men, however, were very different in the way that they approached an inventive task, and this might explain some of their disagreements. Edison, who had limited formal education, often used an intuitive approach, frequently spending hours or days working with different materials—sometimes dozens of them—trying to find one that would work well, or work at all. Sprague, on the other hand, benefited from his scientific training and worked in a much more directed process, proceding in a focused way toward specific designs.

Sprague worked according to a set of proven principles, and regularly used mathematics to help resolve questions or achieve the proper design of an instrument.

Edison was ten years older than Sprague, and had already begun developing the extensive system of workshops, laboratories, and crew of men at Menlo Park, New Jersey, that helped make him the extraordinary inventor of his time. Sprague had made good use of all the technical knowledge that he had learned from the Naval Academy and his innovative mind was soon applying new ideas to electrical needs. That first encounter with Edison was a helpful one for the young Sprague and one of many cordial ones between the two electricians that would follow.

Early in 1882 the navy had sent Sprague to the Electrical Exhibition at the Crystal Palace, near London, where he saw Edison's extensive exhibit—"there was nothing which so impressed me as Edison's work," he later wrote—and had met with Edward H. Johnson, who was a close associate with Thomas Edison, and who was then managing both the Edison installation at the Crystal Palace and the formation of the new Edison Electric Light Co. in England. Johnson had been very impressed with Sprague's knowledge and ability and his work on the Jury of Awards at the exhibition, and before it was over had persuaded Sprague to resign from the navy and to take a new job with the rapidly growing Edison organization.

Sprague's new career as an Edison employee was short-lived. Unhappy with the work he was doing, he soon left the Edison firm to pursue his own interests in electric motors. But if there had ever been any ill feelings about the Sprague-Edison break up, it did not last for long. Sprague's work on his electric motors progressed rapidly, and by the opening of the International Electrical Exhibit in Philadelphia in 1884 he was ready to show his new designs. Even Thomas Edison commented on how well Sprague's motors worked and urged the use of Sprague motors as the "only practical and economic Motor existing today."[1]

Sprague had established his new Sprague Electric Railway & Motor Co. late in 1884, but for the next several years almost all of his electrical manufacturing work was subcontracted to the Edison Machine Works. Sprague's motor business was doing well and the successful electrification of the Richmond, Virginia, street railway system in 1888 created a booming business for Sprague. By 1889 some two-thirds of the electrical manufacturing work of the Edison plant was devoted to manufacturing motors and electrical equipment for Sprague, and the newly established Edison General Electric Co. decided to acquire the Sprague firm.

The aftermath of the sale of the Sprague firm to Edison General Electric led to some bitter disappointment for Frank Sprague that would never be completely overcome in his relationship with Edison. Sprague continued as vice president of the firm until it was merged with Edison General Electric in July 1890, and then continued as a consultant. But Sprague received no workshop or laboratory on which experimental work could be developed, and a proposed new design for an Edison system would be developed by others at Menlo Park, using a low voltage DC current supplied from the rails, a system Sprague felt would be inferior to his design.

Sprague felt, too, that Edison General Electric men were actively disparaging Sprague equipment. The Edison chief engineer, claimed Sprague, had told equipment suppliers that the only guarantee they would give on Sprague equipment was as "a scrap heap of copper and iron." Sprague was typically forceful in stating his opinion, and the Edison men often talked of his "caustic personality." Samuel Insull, who had become a strong right hand to Edison, was often seen as particularly antagonistic toward Sprague.

Sprague would be bitterly disappointed, too, with Edison's early move to replace the Sprague name with "Edison" equipment, or the "Edison System of Electric Railways" on everything developed by the Sprague Electric Railway & Motor Co. In a rather bitter letter of December 2, 1890, resigning his consulting appointment, Sprague spoke of Edison's preference to substitute his own name rather than celebrating the original inventor's accomplishment:

> Instead, it contents itself with the promulgation of circulars known by every railway man in this country to be untrue, and has set out to do everything possible to wipe out the Sprague name, and to give to Mr. Edison the reputation properly belonging to other men's work. No statement is too strong, no reputation as to ability, attainments or work done so false but that it finds ready promulgation. He finds most favor who is most abusive of all things Sprague, and he meets with a cool reputation who does him smallest reverence. The Edison fetish must be upheld, the Sprague name must be abolished; that is the law.[2]

This would prove to be a recurring problem for Frank Sprague when his successful inventions were later sold to other companies. When Sprague decided to get out of the elevator business and sold his Sprague Electric Elevator Co. in 1898, the Sprague name quickly disappeared into the surviving Otis Elevator Co. But with his sale of his Sprague Electric Co. and the multiple unit control system to General Electric in 1902, Sprague had the foresight to include contractual requirements that all equipment be marked "Sprague" or

"Sprague General Electric." Much later, even with this requirement, GE arbitrarily dropped the Sprague marking on its multiple unit equipment in 1928[3] and restored it only after a long letter from Sprague to GE president Gerard Swope.[4]

As Sprague well knew, without the inventor's name being attached to the equipment he invented, the inventor's name is soon forgotten. Consider two of Frank Sprague's contemporary inventors, Edison's incandescent light or Alexander Graham Bell's telephone. Both of them were successful in linking their names to their inventions, and both remain well known even today, while Sprague's name, whose principal inventions were certainly of equal merit, is now largely forgotten in popular history

Sprague soon found himself, too, competing for recognition against the widely exaggerated accounts of Edison's electric railway work. Edison's early work had included the construction of two experimental electric railways in Menlo Park in 1880 and 1881, and he had joined with Stephen D. Field, an electrical engineer and underseas cable inventor, to develop an experimental locomotive that was operated at a Chicago exposition in 1883. These efforts had incorporated neither significant new development, nor were they followed by any other work, and there were few Edison patents on electric railways. Instead, the work of the electric railway went largely to such electrical inventors and engineers as Frank Sprague, Leo Daft, John Henry, Edward Bentley, Walter Knight, Sidney Short, and Charles Van Depoele. Edison's failure to continue with electric railways was due simply to his involvement with the development of electric illumination. "I could not go on with it because I had not time," Edison told *Electrical World*. "I had too many other things to attend to, especially in connection with electric lighting."[5]

But with the success and rapid growth of electric railways, Edison's admirers, careless newspaper reporters, textbook writers, and Edison himself soon began to portray his work on electric railways in greatly exaggerated terms. Roger Burlingame in *Engines of Democracy: Inventions and Society in Mature America*, wrote:

> Genius of his sort is not immune to age. He experienced many failures during his later period. The mythology, founded in great achievement, went on, however, gaining momentum and falsehood with every passing year. He was honored and decorated, a sense of his greatness was forced upon him, he accepted it with reluctance, but no man can resist such pressure. Often when he was given credit for something like electric traction which belonged, *in toto*, to others he seems simply to have refused comment.[6]

And the longer the exaggerations were repeated, the harder it would ever be to correct them.

It was clear that it was extremely important to Frank Sprague that his name be linked to his accomplishments, and for the rest of his life he regularly fired off letters to anyone who had incorrectly credited his work or failed to properly recognize it. Among many such complaints were two particularly notable exchanges that were spread out in the New York newspapers in the decade after the Great War.

The first of these was a long article for the New York *The Sun* on August 10, 1919, by writer Edward Marshall who in an interview with Edison wrote about his important development of the electric trolley car and the mechanism of the "first trolley car" in Orange, New Jersey. "It was the pioneer appliance of cheap, quick power street transportation in America," wrote Marshall. "Naturally he is proud of it."[7]

Two weeks later Sprague's rejoinder to this exaggeration was headed by a long paper to *The Sun* titled, "Inventors of the Electric Railway: Frank J. Sprague Protests Against the Share of the Glory Assigned in an Interview to Thomas A. Edison," in which Sprague spelled out Edison's limited involvement in electric railways, and his own and others' much greater work to develop the successful trolley car.[8] Over the next several weeks Edison[9] and Sprague[10] in a series of headlines *The Sun* called, "The Great Men of the Electrical Age," exchanged further differences of opinion on each one's work on the trolley car.

In recalling the experience of the Richmond trolley car project, Edison reminded Sprague that his Edison Machine Works—which manufactured most of Sprague's electrical equipment—had trusted Sprague for more then $100,000 of apparatus when the Richmond project, as Edison put it, "was in a serious situation, due to too much mathematics on the part of Mr. Sprague."

A similar uproar followed an article by James C. Young, "The Magic Edison Made for the World," published on August 12, 1928, in *The New York Times Magazine*. "Within one decade," Young told us, "Edison accomplished three things at Menlo Park—the phonograph, the incandescent light and the electric car," and went on to describe how Edison had done it all.[11]

The following month, Frank Sprague was back with a long rebuttal of the exaggerated Edison claims, with a detailed account of the work of the many others, as well as his own, that had made electric transportation possible. "Electric Railway Not the Creation of One Man," said the *Times*.[12]

But by this time it had become much more difficult to revise accounts of the establishment of electric railways to put them on a more factual basis for the man who was often called "our greatest inventor." Already published over the years had been far too many popular newspaper accounts, school books and the like that in some manner had misstated or exaggerated the story of electric traction.

In March 1928, for example, Frank Sprague had written to the Congress, urging that papers concerning Edison's work accompanying the award of a Congressional Medal of Honor to him be corrected before the planned award. Congressman Randolph Perkins, chairman of the medal committee, replied to Sprague: "No one can write of Mr. Edison without overwriting. Moreover, I cannot understand how there could possibly be any criticism of anything said kindly about Mr. Edison. Surely, he deserves every sort of exaggeration."[13]

Even today the overstatement of Edison's work on the electric railway sometimes continues. Still in print, for example, is the children's book published by Landmark Books and written by Margaret Cousins, *The Story of Thomas Alva Edison*.[14] The author discusses Edison's interest in electric transportation, gives a brief description of his experimental electric railways at Menlo Park in 1880 and 1881, and then without any mention of Frank Sprague, Leo Daft, John Henry, Edward Bentley, Walter Knight, Sidney Short, Charles Van Depoele, or other inventors of the street railway, leaps across the years to the early 1900s to credit Edison with electrification of the New York Central Railroad, with which Edison had nothing to do at all! The youthful reader is likely to suppose that Edison had done it all.

While several Edison biographers and other historical writers have touched upon the many examples of exaggerated claims for Edison's accomplishments, few have ever taken the subject on directly, or in detail. Frank Sprague had long planned to develop a history of electrification which would, among other things, clearly establish the accomplishments of the many pioneers, but he never could take enough time away from his electrical pursuits to do it. However his widow, Harriet Sprague, took on the task in her monograph, *Frank J. Sprague and the Edison Myth*, in 1947. She did it well and in fairness, closing her words in this way:

> In Memory of two great inventors and of the many others whose inventive genius helped to create our present civilization and the industrial age, let us hope that it may now be possible to close the books on all controversy and offer deserved tribute to each.[15]

Apart from their electrical work, Sprague and Edison had serious differences during their tenure on the Naval Consulting Board. But both men also regarded the other as a remarkably accomplished inventor and electrical engineer. Edison, not usually loud in his praises of other engineers and inventors, was most generous in his public endorsement of Sprague's early work on electric motors, and even in the midst of a public quarrel with Sprague in the pages of New York's *The Sun* in 1919, Edison maintained: "What Mr. Sprague can really claim is that by his persistence and electrical knowledge he was the first one to start the trolley car system commercially. This is what I have always thought and publicly stated."[16]

Sprague, in his turn, also held Edison's work in electric illumination and other areas in high regard. In a December 1931 letter to *Electrical Engineering* magazine shortly after Edison's death in October 1931, Sprague maintained that Edison's greatest work was electrical incandescence, stating in part:

> "A filament of carbon of high resistance" enclosed in a vacuum bulb hermetically sealed—the basis of his first demonstration in 1879 and of the foundation patent in 1880—with the collateral central station developments, around these will finally be written the Edison saga, for, regardless of the importance of his other inventions, the development and mastery of electric lighting by incandescence was Mr. Edison's most dramatic and difficult work, and its influence upon the modern electric age his most far-reaching one.
>
> But whatever the developments in the electric art and however wide departures in practice, the vital fact remains that Mr. Edison established the first real control station for the general distribution of electricity, and his name has become a symbol of fertility of invention and untiring activity, with results which have made the world forever his debtor. He has indeed lighted the world.[17]

SPRAGUE THE INVENTOR ON SPRAGUE THE INVENTOR

We often are cautioned that we should "know ourselves," and equally as often cautioned that such knowledge is elusive. It is unlikely that anybody who knew Frank Sprague personally (and certainly no one who knows him merely from what he wrote of himself) would have described him as having false modesty. Sprague was proud of his accomplishments, and not shy of saying so, as often as he felt necessary. He was equally as certain that his solution to a given problem or challenge was the best possible. However, Sprague's sense of himself was not based in braggadocio or egotism; it was based on a sense of what he believed to be right, which included his opinion of his solutions to specific design challenges.

Sprague's sense of his own rightness is manifest in his many statements regarding the multiple unit system. He repeatedly describes resistance to the system as "obstructionism," and one can sense his frustration that others failed to see the merit of his inventions as clearly as he did. In this spirit, he worked tirelessly to promote his inventions, referring to this as his "missionary work." Between 1885 and 1904 he made over 50 demonstrations, lectures, and publications on electric traction.[18] He approached the Manhattan elevated at least six times with a variety of proposals, including that he should undertake the work at his own risk, before they finally acquiesced in 1901 (under contract with General Electric). He could come across as rather smug when he felt that his opinions had been vindicated and his proposals adopted.

He also could be dismissive of alternate views, which he found inferior to his own, such as his description of his first encounter with "the Edison system," where he developed a new method for calculating electrical distribution loads.

> 1883: . . . Developed mathematical method of determining sizes and disposition of mains and feeders in system of electrical distribution, as distinguished from erroneous empirical and experimental methods used by Edison and his associates.
>
> Edison's plan had for its basis the idea of a plurality of groups of lights of about equal numbers which although the mains were connected in a net work were each to be supplied by feeders carrying equal loads, and hence to get the same drop of potential of equal resistances,—a serious misconception of the needs and facts.
>
> See Edison patents No. 264,642 and 266,793
>
> Sprague's method, adopted after soundness was demonstrated, was to use mains of varying sizes, to get equal maximum drop between supply points, and to provide feeders of resistance inversely proportional to the normal loads.
>
> Patent #335,045, Jan. 26, 1886; App. For Sept. 19 1885.[19]

Although he could be very prickly on issues of priority or precedence, Sprague never appears to have been eager to claim accolades that he hadn't earned. He also was generous in sharing credit with those who had earned it and in pointing out the contributions of others (and Edison's failure to do likewise clearly irked him to no end). His view of his own accomplishments, however, could be decidedly one-sided. His previously cited statement on his dispute with Van Depoele over the invention of the trolley pole is a good example. Although he claimed that the decision awarding the patent to Van Depoele was unjust, it is fairly clear, even from his own description of his earlier work, that the only element of his earlier design in which he had really anticipated Van Depoele was the wheel which made contact with the current source. In the case of the 1885 patent application, it was an overhead rail, not a wire. At the time he was working on this patent (1882–1883), he was thinking of the London Underground, in which a conducting rail would be mounted on the tunnel's roof.[20]

Overall, Sprague's writings give us the impression of an inventor who, although usually quite fair in his dealings with others, was powerfully driven by the conviction that he had arrived at the superior solution. When faced with "obstructionism," or worse, if his designs were slighted, ignored, or accredited to someone else he could turn quite abrasive. He clearly was a man with a strong sense of the significance of his own work.

THE FRANK SPRAGUE LEGACY

From the time Sprague's first patent was approved by the Patent Office on March 18, 1884, for an "Electro-Dynamic Motor," until his last one was approved 48 years later on April 20, 1932, for a "method of and apparatus for control of train movements," he had successfully obtained 95 patents, covering such diverse topics as electric motors, electric railway systems, multiple unit train controls, automatic train controls, electric elevator controls, moving displays and signs, and such miscellaneous topics as a system for differential air pressure on ships or an electric bread toaster.

In assessing the value of any inventor's work, much more important than the number of patents is an assessment of how widely it was adopted, what value it had for its users, or for how long its use was continued. It is in the nature of inventions that a great many of the ideas successfully patented never get used at all; no one sees a need for it, or someone else has a different or better solution to the need.

In looking at Frank Sprague's work it is interesting to note that, while he continued his inventive work until the time of his death at age 77, almost all of his most successful inventions were completed by the time he had reached 40. Such later work as the automatic train controls work, to which Sprague devoted a quarter century of effort from 1907 to almost the end of his life, was by all accounts an extremely well engineered system, but for whatever reason, almost all of the automatic train control work for railroads was sold by other manufacturers. The innovative dual elevator system Sprague completed in 1931 won wide public mention, but at the start of the Great Depression no one needed an elevator for taller new buildings, and only a single test installation was ever made. His Universal Electric Sign System was still under development when Sprague died.

The most successful Sprague inventions fell into four groups, most of them preceding 1900. Many of his early electric motor inventions helped develop the electric power system of the 1880s, while Sprague's electric elevator controls of the 1890s led the industry from hydraulic and steam-powered to electric systems.

By far the largest group of Sprague patents was devoted to the electric railway systems which he developed primarily through the 1880s. Sprague worked among a number of other competing inventors, all trying to establish a workable street railway system. While almost every inventor had contributed one or more useful features to the electric railway, Sprague perfected several key components and brought to it the engineering work required to make it a commercially feasible system in Richmond. Sprague's work made him first of several among electric railway inventors and earned him the title of "the father of electric traction."

Filling an important need for growing cities, the street railway system developed into a major industry in North America and for most of the world over the next several decades. Based upon the designs of Sprague or other inventors, the street railway flourished for more than a half century before largely being displaced in urban transit by motor bus services in the mid-twentieth century.

But Frank Sprague's street railway has demonstrated a remarkably useful longevity, and today is being revived in what might be called its "second evolution," more than a century after its original development following Sprague's success in Richmond. A few of these original street railways survived, and urban planners began to appreciate the superior ability of the street railway to handle large crowds of urban travelers. Original street railway lines have been rebuilt and modernized, and many more new ones have been constructed in lines that are usually called "light rail lines." While these new electric cars incorporate power systems and controls much advanced from those installed in the original equipment, they provide the same urban transportation services. And the basic track and power supply—with a standard street track and a 600-volt DC overhead power supply—are little changed from those of their nineteenth century prototypes. On some lines modern light rail cars even operate together with what are usually called "heritage" lines, using rebuilt cars almost a hundred years old, or modern replicas of early twentieth-century trolley cars.

But above all others was Sprague's inventive work on the multiple unit system. Unlike so many discoveries which involved multiple inventors working

The regional Sound Transit authority began operating this modern streetcar in downtown Tacoma, Washington, in 2003, while new light rail lines in Seattle are scheduled to begin operating in 2009. *William D. Middleton.*

toward the same end point, Frank Sprague's invention of the multiple unit system control was his alone. When his system was proposed for the South Side road's electrification in 1897 his was the only one available and his competitors could only offer the then-standard locomotive-hauled equipment. Within less than a decade New York and Chicago's elevated and subway systems had all converted to multiple unit control, and the system was soon the standard all over the world. Today almost a hundred urban rapid transit systems are operated on a multiple unit control on every continent except Antarctica, and many more are operated on light rail systems and railroads on a worldwide basis. Of all of Frank Sprague's many inventions, none have served the world more broadly or as long, or has provided such significant benefits as his invention of multiple unit system control.

A writer for the *New York Herald Tribune* summarized Frank Sprague's legacy two days after his death in 1934. He had it right:

More than a hundred years after Frank Sprague put his first multiple unit control system in service on the Chicago "L," the system still serves Chicago rapid transit—and close to a hundred other systems elsewhere around the world—well. This was a southbound train on Chicago's "Loop," at Wabash Avenue and Madison Street in 1987.   *William D. Middleton.*

Father of the Trolley Car

The death of Frank J. Sprague removes the third of a remarkable trio of American inventors who made notable the closing quarter of the last century. Perhaps no three men in human history have done more to change the daily lives of human kind. Certainly no three have been responsible for greater progress in use of electricity. Edison, the telegrapher, made it possible for human beings to work or play at night with cheap electric light. Bell, the singing teacher, took the first steps in that vast development of telephony and radio which now makes it possible for one person almost anywhere in the world to converse with another person almost anywhere else. Sprague, the young naval officer, dreamed of railways and street railways run cheaply and safely by this motive power of electricity and made his dream come true. All three men might stand as models of the peculiar practical genius of America. There were electric lamps before Edison. There were telephones of a sort before Bell. There were electric motors before Sprague. But these three men made these three things work.

A year ago last July Mr. Sprague himself described that incredible adventure in Richmond in 1887 which resulted in the first electric street railway. Promoters in the Richmond city had obtained a franchise and had constructed something that passed for a track. Only three years out of the navy, young Sprague had a small business building motors. Plans for a street railway had been worked out in his head, but nothing of the kind ever had been built. All experimental cars and other electric vehicles then in existence anywhere did not total the forty streetcars required to start the Richmond line. Yet Sprague and his associates took a contract to supply the cars and electrify that line in ninety days. What is still more remarkable, they did it.

That was the beginning of practical electric transportation, yet Mr. Sprague's fame in future engineering probably will rest less on that than on the development which Mr. Frank Hedley has mentioned as foremost in the long list of Sprague inventions. This is the multiple-control system by which several motor cars of an electric train, such as a subway train, all are operated at once from the cab of a single motorman. In city transportation this is necessary, both to provide trains of any required number of cars and to make possible the rapid starting needed for speed on lines with many station stops. Vertical transportation by electric elevators also owes much to Mr. Sprague's inventions. Not merely as father of the trolley, but of all transportation by electricity, his fame is firm.[21]

APPENDIX A
Frank Julian Sprague's Patents

Frank Sprague was issued 95 patents during his career. They are listed in the order in which they were issued, by number, title, patentees, application date, issue date, and assignee.

TABLE A.1.

Number	Title	Patentees	Application Date	Issue Date	Assignee
295,454	Electro-Dynamic Motor	FJS	May 2, 1883	March 18, 1884	
304,145	Dynamo-Electric Machine	FJS	October 4, 1881	August 26, 1884	
309,167	Adjustable Resistance for Electrical Circuits	FJS	April 2, 1884	December 9, 1884	Assigned to S. Insull
313,247	Regulator For Electro-Dynamic Motors	FJS	February 21, 1885	March 3, 1885	
313,546	Electro-Dynamic Motor	FJS	November 4, 1884	March 10, 1885	Later assigned to Sprague Electric Railway and Motor Company
314,891	Electrical Indicator	FJS	January 24, 1884	March 31 1885	
315,179	Electro-Dynamic Motor	FJS	April 30, 1884	April 7, 1885	
315,180	Electro-Dynamic Motor	FJS	June 9, 1884	April 7, 1885	
315,181	Electric Motor And Generator	FJS	July 19, 1884	April 7, 1885	
315,182	Electric Motor And Generator	FJS	July 19, 1884	April 7, 1885	
315,183	Electro-Dynamic Motor	FJS	January 19, 1885	April 7, 1885	
317,235	Electric Railway	FJS	April 14, 1882	May 5, 1885	Assigned to Sprague Electric Railway and Motor Company
318,668	Method Of Operating Electric-Railway Trains	FJS	December 22, 1884	May 26, 1885	Assigned to Sprague Electric Railway and Motor Company
321,147	Electro-Dynamic Motor	FJS	February 27, 1885	June 30, 1885	Assigned to Sprague Electric Railway and Motor Company
321,148	Electro-Dynamic Motor	FJS	March 3, 1885	June 30, 1885	Assigned to Sprague Electric Railway and Motor Company
321,149	Electric Railway System	FJS	March 21, 1885	June 30, 1885	Assigned to Sprague Electric Railway and Motor Company
321,150	Electro-Dynamic Motor	FJS	May 20, 1885	June 30, 1885	Assigned to Sprague Electric Railway and Motor Company

323,459	Electric-Railway System	FJS	January 19, 1885	August 4, 1885	Assigned to Sprague Electric Railway and Motor Company
323,460	Electro-Dynamic Motor	FJS	March 12, 1885	August 4, 1885	Assigned to Sprague Electric Railway and Motor Company
324,891	Electro-Dynamic Motor	FJS	March 12, 1885	August 25, 1885	Assigned to Sprague Electric Railway and Motor Company
324,892	Electric Railway-Motor	FJS	May 25, 1885	August 25, 1885	
328,821	Electric-Railway System	FJS	August 27, 1885	October 20, 1885	
335,045	System of Electrical Distribution	FJS	September 19, 1885	January 26, 1886	Assigned to The Edison Electric Light Company
335,781	Electro-Dynamic Motor	FJS	September 26, 1885	February 9, 1886	
337,793	Electro-Dynamic Motor	FJS	July 6, 1885	March 9, 1886	
337,794	Electro-Dynamic Motor	FJS	July 6, 1885	March 9, 1886	
338,313	Electric Railway System	FJS	December 15, 1884	March 23, 1886	Assigned to Sprague Electric Railway and Motor Company
338,619	Electric Railway	FJS	April 14, 1882, Application divided May 6, 1885	March 23, 1886	
340,684	Electric Railway	FJS	November 24, 1885	April 27, 1886	
340,685	Electric Railway	FJS	November 24, 1885	April 27, 1886	
353,829	Electrical Propulsion of Vehicles	FJS	June 12, 1886	December 7, 1886	
372,822	Dynamo-Electric Machine	FJS	March 3, 1885	November 8, 1886	Assigned to Sprague Electric Railway and Motor Company
372,823	Electric Motor	FJS	April 18, 1887	November 8, 1886	
372,824	Electrical Pumping Apparatus	FJS	June 4, 1887	November 8, 1886	
380,144	Dynamo-Electric Machinery	FJS	April 8, 1887	March 27, 1888	Assigned to Sprague Electric Railway and Motor Company

Number	Title	Patentees	Application Date	Issue Date	Assignee
385,211	Electrical Pumping Apparatus	FJS	June 4, 1887	June 26, 1886	
387,745	Gearing	FJS	April 18, 1887	August 14, 1887	
397,875	Overhead Line for Electric Railways	FJS	September 27, 1888	February 12, 1889	Assigned to Sprague Electric Railway and Motor Company
406,600	Electric Railway-Motor	FJS	March 1, 1889, Application divided June 4, 1889	July 9, 1889	Assigned to Sprague Electric Railway and Motor Company
408,544	Electric-Railway System	FJS	February 27, 1885	August 6, 1889	Assigned to Sprague Electric Railway and Motor Company
413,151	Electric-Railway Car	FJS	July 5, 1889	October 15, 1889	Assigned to Sprague Electric Railway and Motor Company
414,172	Electric Railway	FJS, Patrick F. O'Shauhnessy	January 22, 1889, Application divided September 7, 1889	October 29, 1889	Assigned to Sprague Electric Railway and Motor Company
424,436	Electric-Railway Motor	FJS	March 30, 1889	March 25, 1890	Assigned to Sprague Electric Railway and Motor Company
428,732	Electric Motor and Generator	FJS	July 19, 1884	May 27, 1890	Assigned to Sprague Electric Railway and Motor Company
429,327	Electric-Railway Motor	FJS	October 31, 1889	June 3, 1890	Assigned to Sprague Electric Railway and Motor Company
431,823	Electric-Railway Motor	FJS	June 11, 1889	July 8, 1890	Assigned to Sprague Electric Railway and Motor Company
433,425	Contact Device For Electric Railways	FJS	July 2, 1889	July 29, 1890	Assigned to Sprague Electric Railway and Motor Company
438,293	Electric Railway System	FJS	December 15, 1884, Application divided May 21, 1885	October 14, 1890	

Patent No.	Title	Inventors	Filed	Granted	Assignment
445,515	Conductor-Switch For Electric Railways	FJS, Johan F. S. Branth	September 27, 1888	January 27, 1891	Assigned to Sprague Electric Railway and Motor Company
461,552	Field-Magnet For Dynamo-Electric Machines	FJS	January 19, 1885	October 20, 1891	Assigned to Sprague Electric Railway and Motor Company
465,218	Adjustable Rheostat	FJS, Charles R. Pratt	August 25, 1891	December 15, 1891	
465,806	Electric-Railway Trolley	FJS, Patrick F. O'Shauhnessy	January 22, 1889	December 22, 1891	Assigned to Sprague Electric Railway and Motor Company
466,448	Communicator-Brush And Holder For Dynamo-Electric Machines and Motors	FJS	December 10, 1886	January 5, 1892	Assigned to The Edison General Electric Company
468,959	Electric Railway	FJS, Patrick F. O'Shauhnessy	January 19, 1889	February 16, 1892	Assigned to Sprague Electric Railway and Motor Company
481,739	Electric Motor And Regulating Device Thereof	FJS	April 29, 1889	August 30, 1892	Assigned to Sprague Electric Railway and Motor Company
504,255	Electric-Railway Trolley	FJS, Patrick F. O'Shauhnessy	January 19, 1889	August 29, 1893	Assigned to Sprague Electric Railway and Motor Company
637,175	Automatic Damper and Valve Regulator	FJS	April 3, 1899	November 14, 1899	Assigned to The Howard Thermostat Company
647,239	Elevator	FJS	July 27, 1898	April 10, 1900	Assigned to Sprague Electric Company
647,240	Electrically-Driven Mechanism	FJS	July 27, 1898, Application divided September 20, 1899	April 10, 1900	Assigned to Sprague Electric Company
647,241	Brake For Hoisting Mechanism	FJS	July 27, 1898, Application divided September 20, 1899	April 10, 1900	Assigned to Sprague Electric Company
647,242	Cable-Winding Safety Device	FJS	July 27, 1898, Application divided September 20, 1899	April 10, 1900	Assigned to Sprague Electric Company

Number	Title	Patentees	Application Date	Issue Date	Assignee
652,049	Railway Car	FJS	December 30, 1898	June 19, 1900	Assigned to Sprague Electric Company
660,065	Traction System	FJS	April 30, 1898	October 16, 1900	Assigned to Sprague Electric Company
660,066	Method Of Regulating Motors	FJS	April 30, 1898, Application divided June 23, 1900	October 16, 1900	Assigned to Sprague Electric Company
696,880	Traction System	FJS	December 16, 1898	April 1, 1902	Assigned to Sprague Electric Company
716,953	Electric Elevator	FJS	October 29, 1894	December 30, 1902	Assigned to Sprague Electric Company
765,686	Electrically-Controlled Damper for Heaters	FJS	September 4, 1903	July 26, 1904	
774,611	Electrical Coupling	FJS	April 30, 1898, Application divided July 7, 1900	November 8, 1904	Assigned to Sprague Electric Company
815,756	System of Electrical Control	FJS	September 22, 1898	March 20, 1906	Assigned to Sprague Electric Company
908,180	Third Rail	William J. Wilgus, FJS	May 27, 1905	December 29, 1908	
1,113,257	Method and Means for Automatically Applying Differential Air-Pressure to Compartments of Ships	FJS, Frank D. Sprague	March 15, 1912	October 13, 1914	
1,248,942	Electric Detector Circuit	FJS	December 31, 1914, Application divided February 26, 1916	December 4, 1917	Assigned to Sprague Safety Control and Signal Corporation
1,322,148	Recorder	FJS	December 31, 1914, Application divided February 26, 1916	November 18, 1919	Assigned to Sprague Safety Control and Signal Corporation

Patent Number	Title	Inventor	Application Date	Issue Date	Assignment
1,553,295	Pressure Equalizer for Air-Brake Systems	FJS	December 31, 1914, Application divided June 3, 1925	September 8, 1925	Assigned to Sprague Safety Control and Signal Corporation
1,581,094	Train-Control Apparatus	FJS	December 31, 1914, Application divided September 30, 1925	April 13, 1926	Assigned to Sprague Safety Control and Signal Corporation
1,669,265	Automatic Train-Control System	FJS	January 7, 1922	May 8, 1928	Assigned to Sprague Safety Control and Signal Corporation
1,706,259	Electric Toaster	FJS	October 5, 1926	March 19, 1929	
1,742,421	Train-Control Apparatus and Circuit Arrangement	FJS	August 7, 1919, Application divided May 10, 1927	January 7, 1930	Assigned to Sprague Safety Control and Signal Corporation
1,763,198	Dual Elevator System and Control	FJS	December 31, 1926	June 10, 1930	Assigned to Westinghouse Electric & Manufacturing Company
1,771,149	Apparatus for Controlling the Movement of Vehicles on Railroads	FJS	April 12, 1927, Application divided July 6, 1929	July 22, 1930	
1,780,148	Method and Apparatus for Control of Train Movements	FJS	December 31, 1914	October 26, 1930	Assigned to Sprague Safety Control and Signal Corporation
1,822,633	Electric Toaster	FJS	April 13, 1929	September 8, 1931	
1,835,912	Universal Electric Sign System	FJS	November 19, 1929	December 8, 1931	Assigned to Sprague Signs, Inc.
1,852,487	Device for Compressing Corrugated Tubes	FJS, Max Knabb	September 5, 1930	April 5, 1932	Assigned to Sprague Specialties Company
1,860,908	Apparatus for Controlling the Movement of Vehicles on Railroads	FJS	April 12, 1927, Application divided July 6, 1929	May 31, 1932	Assigned to Sprague Safety Control and Signal Corporation
1,869,627	Apparatus for Controlling the Movement of Vehicles on Railroads	FJS	April 12, 1927, Application divided July 6, 1929	August 2, 1932	Assigned to Sprague Safety Control and Signal Corporation

Number	Title	Patentees	Application Date	Issue Date	Assignee
1,876,032	Apparatus for Controlling the Movement of Vehicles on Railroads	FJS	April 12, 1927	September 6, 1932	Assigned to Sprague Safety Control and Signal Corporation
1,886,062	Train Control	FJS	June 12, 1928	November 1, 1932	
1,889,724	Method of and Apparatus for Control of Train Movements	FJS	August 7, 1919	November 29, 1932	Assigned to Sprague Safety Control and Signal Corporation
1,971,281	Display Device	FJS	January 30, 1933	August 21, 1934	
1,978,966	Apparatus for the Manufacture of Perforated Records	FJS	July 6, 1931	October 30, 1934	Assigned to Sprague Signs, Inc.
1,978,967	Electric Selector Device	FJS	July 6, 1931, Application divided January 20, 1933	October 30, 1934	Assigned to Sprague Signs, Inc.
1,987,719	Safety Mechanism for Elevators	FJS	October 10, 1928	January 15, 1935	Assigned to Westinghouse Electric & Manufacturing Company
2,001,163	Train Control Apparatus	FJS	December 31, 1914, Application divided August 24, 1927	May 14, 1935	Assigned to General Railway Signal Company
2,024,074	Universal Electric Sign System	FJS	November 19, 1929	December 10, 1935	Assigned to Sprague Signs, Inc.

SPRAGUE PATENTS BY CATEGORY

Elevators

647,239	Elevator
647,241	Brake For Hoisting Mechanism
647,242	Cable-Winding Safety Device
716,953	Electric Elevator
1,763,198	Dual Elevator System and Control
1,971,281	Display Device
1,987,719	Safety Mechanism for Elevators

Motors

295,454	Electro-Dynamic Motor
304,145	Dynamo-Electric Machine
313,247	Regulator For Electro-Dynamic Motors
313,546	Electro-Dynamic Motor
315,179	Electro-Dynamic Motor
315,180	Electro-Dynamic Motor
315,181	Electric Motor And Generator
315,182	Electric Motor And Generator
315,183	Electro-Dynamic Motor
321,147	Electro-Dynamic Motor
321,148	Electro-Dynamic Motor
321,150	Electro-Dynamic Motor
323,460	Electro-Dynamic Motor
324,891	Electro-Dynamic Motor
335,781	Electro-Dynamic Motor
337,793	Electro-Dynamic Motor
337,794	Electro-Dynamic Motor
372,822	Dynamo-Electric Machine
372,823	Electric Motor
380,144	Dynamo-Electric Machinery
428,732	Electric Motor And Generator
461,552	Field-Magnet For Dynamo-Electric Machines
466,448	Communicator-Brush And Holder For Dynamo-Electric Machines and Motors
481,739	Electric Motor And Regulating Device Thereof
660,066	Method Of Regulating Motors

Other

372,824	Electrical Pumping Apparatus
385,211	Electrical Pumping Apparatus
637,175	Automatic Damper and Valve Regulator
647,240	Electrically-Driven Mechanism
765,686	Electrically-Controlled Damper for Heaters
1,113,257	Method and Means for Automatically Applying Differential Air-Pressure to Compartments of Ships
1,322,148	Recorder
1,706,259	Electric Toaster
1,822,633	Electric Toaster
1,835,912	Universal Electric Sign System
1,852,487	Device for Compressing Corrugated Tubes
1,978,966	Apparatus for the Manufacture of Perforated Records
1,978,967	Electric Selector Device
2,024,074	Universal Electric Sign System

Power

309,167	Adjustable Resistance for Electrical Circuits
314,891	Electrical Indicator
335,045	System of Electrical Distribution
465,218	Adjustable Rheostat
774,611	Electrical Coupling
815,756	System of Electrical Control

Traction		**Train Control**	
317,235	Electric Railway	1,248,942	Electric Detector Circuit
318,668	Method Of Operating Electric-Railway Trains	1,553,295	Pressure Equalizer for Air-Brake Systems
321,149	Electric Railway System	1,581,094	Train-Control Apparatus
323,459	Electric-Railway System	1,669,265	Automatic Train-Control System
324,892	Electric Railway-Motor	1,742,421	Train-Control Apparatus and Circuit Arrangement
328,821	Electric-Railway System		
338,313	Electric Railway System	1,771,149	Apparatus for Controlling the Movement of Vehicles on Railroads
338,619	Electric Railway		
340,684	Electric Railway		
340,685	Electric Railway	1,780,148	Method and Apparatus for Control of Train Movements
353,829	Electrical Propulsion of Vehicles		
387,745	Gearing	1,860,908	Apparatus for Controlling the Movement of Vehicles on Railroads
397,875	Overhead Line for Electric Railways		
406,600	Electric Railway-Motor	1,869,627	Apparatus for Controlling the Movement of Vehicles on Railroads
408,544	Electric-Railway System		
413,151	Electric-Railway Car		
414,172	Electric Railway	1,876,032	Apparatus for Controlling the Movement of Vehicles on Railroads
424,436	Electric-Railway Motor		
429,327	Electric-Railway Motor		
431,823	Electric-Railway Motor	1,886,062	Train Control
433,425	Contact Device For Electric Railways	1,889,724	Method of and Apparatus for Control of Train Movements
438,293	Electric Railway System	2,001,163	Train Control Apparatus
445,515	Conductor-Switch For Electric Railways		
465,806	Electric-Railway Trolley		
468,959	Electric Railway		
504,255	Electric-Railway Trolley		
652,049	Railway Car		
660,065	Traction System		
696,880	Traction System		
908,180	Third Rail		

APPENDIX B
Frank Julian Sprague Awards and Honors

Awards

Gold Medal, Paris Exposition, 1889, "presented to the Sprague Electric Railway and Motor Company for the most perfect system of Electric Rail Way Equipment."

Elliott Cresson Medal, The Franklin Institute, 1904, for the development of the multiple unit system of electric traction.

Grand Prize, Louisiana Purchase Exposition, St. Louis, 1904, for Invention and Development on Electric Railways.

Edison Medal, American Institute of Electrical Engineers, 1910, "for Meritorious Achievement in Electrical Science, Electrical Engineering, or the Electrical Arts."

The Franklin Medal, The Franklin Institute, 1921, "in recognition of his many and fundamentally important inventions and achievements in the field of electrical engineering; notably contributions to the development of the electric motor and its application to industrial purposes, and in the art of electric traction, signally important in forming the basis of world-wide industries and promoting human welfare."

Frank Julian Sprague Bronze Portrait Bust, by sculptor Florence M. Darnault, was developed by Frank Hedley, president of the Interborough Rapid Transit Co., on behalf of the Frank J. Sprague Anniversary Committee, and presented to Sprague at the January 1934 annual meeting of the American Institute of Electrical Engineers in honor of his achievements in electrical engineering and invention.

John Fritz Medal from representatives of four national engineering societies, the American Society of Civil Engineers, American Institute of Mining and Metallurgical Engineers, The American Society of Mechanical Engineers, and the American Institute of Electrical Engineers, 1935, "for distinguished service as inventor and engineer through the application and control of electric power in transportation systems."

Honorary Degrees

Doctor of Engineering (D. Eng.), Stevens Institute of Technology, 1921, was cited at Stevens' 49th annual commencement with a summary of his long and diverse work and "for electrical inventions and achievements forming basis of world industries and promoting human welfare."

Doctor of Science (D. Sci.), Columbia University, 1922, who was cited as follows: "Graduated at the United States Naval Academy with the Class of 1878; early joining the ranks of leadership in invention and discovery in the new field of electrical science; pioneer in the development of the electric railway; long since possessed of a well deserved and world-wide reputation for mastery of the problems of electrical engineering."

Doctor of Laws (LL.D.), University of Pennsylvania, 1924, who was cited as "Pioneer in railway electrification; Equipped first modern trolley road at Richmond, Va., 887; Inaugurated high speed and house automatic elevators; Invented multiple unit system of electronic train control and promoted high tension direct current system; Developed system of automatic signal and brake control; Promoter of underground rapid transit; Awarded Grand Prize, Paris, 1889, for electric railway development; Awarded Grand Prize, St. Louis, 1904, for invention and development in electric railways; Awarded Franklin Medal, 1921, and many others.

"In recognition of your services to the world I confer upon you the Honorary Degree of Laws."

Distinguished Offices and Memberships

President, American Institute of Electrical Engineers, 1892–1893.

Member, Board of Visitors, U.S. Naval Academy, 1906.

Honorary Membership, The Franklin Institute, 1921.

Honorary Membership, American Institute of Electrical Engineers, 1932.

President, New York Electrical Society.

President, American Institute of Consulting Engineers.

President, Inventors' Guild.

APPENDIX C
Common Electrical Terms

Electric Terms

Electric Current, symbol I, is the flow of electricity through a conductor.

Electric Resistance, symbol R, is that which opposes the transmission of electric current in material. Materials which are relatively small in resistance are called conductors; those of such high resistance that they can practically suppress electric transmission are called non-conductors or insulators.

Electromotive Force, symbol emf or E, is that electric condition resulting from difference of potential by which electricity is transmitted from points having positive potential to those having negative potential, often called electric pressure.

Ohm's Law, discovered by German professor George Simon Ohm (1787–1854), states that the current in an electric circuit is directly proportional to the electromotive force in the circuits. These may be expressed by the formula:

$I = \frac{E}{R}$, from which the values of E and R may be derived, $E = IR$, and $R = \frac{E}{I}$

Potential is for electric current what "head" or fall is to currents of water. Electricity flows from points of higher potential, to points of lower potential, or from positive to negative.

Electrical Units

Ampere, sometimes abbreviated amp, symbol A, is the unit of electric current that will produce a defined force defined by an international SI standard. It was named in honor of French physicist André-Marie Ampère (1775–1836).

Electric Horsepower is the electric equivalent of the mechanical horsepower which is called the electric horsepower, abbreviated hp, and is represented by 746 watts.

Ohm, symbol Ω, is the unit of electric resistance, with the standard value for one ohm defined by an international SI standard. It was named in honor of Professor George Ohm.

Volt, symbol V, is the unit of electromotive force, or electric pressure, with the standard value for one volt defined by an international SI (for Le Système International d'Unités) standard. It was named in honor of Italian professor Alessandro Volta (1745–1827).

Watt, symbol W, is the energy by which electric work is performed, with one watt defined as the power expended when one ampere flows between two points having a potential difference of one volt, and is incorporated into the international SI standard. It was named in honor of Scottish engineer and inventor James Watt (1736–1819).

NOTES

Foreword

1. Harriet Sprague, *Frank J. Sprague and the Edison Myth*, (New York: William-Frederick Press, 1947).

2. Frank Rowsome, *Trolley car treasury: A Century of American Streetcars, Horsecars, Cable Cars, Interurbans, and Trolleys* (New York: McGraw-Hill, 1956).

3. Harold C. Passer, *Frank Julian Sprague, Father of Electric Traction, 1857–1934* (Cambridge, Mass.: Harvard University Press, 1952; reprinted from William Miller, ed., *Men in Business* (Cambridge, Mass.: Harvard University Press; 1952).

1. A Boyhood in New England

1. Letter to Sprague Descendents, June 1, 1912, Frank Julian Sprague Papers, Mss. & Archives Section, N.Y.P.L.

2. "Memoir of Frank J. Sprague," *Transactions of the American Society of Civil Engineers*, Vol. 100 (1935): 1736–1741.

3. Ibid., 1737.

4. North Adams, while a separate area of industrial and commercial activity several miles north of the Adams town center, remained part of Adams until the two areas were separated in 1878.

5. Frank J. Sprague, remarks at 75th birthday celebration, in *Frank J. Sprague: Seventy-fifth Anniversary, July 25, 1932* (New York), 29.

6. North Adams, Massachusetts, *Transcript*, Aug. 1885.

7. Ibid.

8. Frank J. Sprague, remarks at 75th birthday celebration, in *Frank J. Sprague: Seventy-fifth Anniversary*, ibid.

9. Letter from Isaac W. Dunham, Superintendent of Public School of North Adams, Mass., June 30, 1874, From The National Archives of the United States.

10. Letter to Frank Sprague from D. C. Sprague, July 9, 1874, Frank Julian Sprague Papers, Mss. & Archives Section, N.Y.P.L.

11. Frank J. Sprague, *Proceedings of the American Electric Railway Association*, Aug. 1932.

2. The Midshipman Inventor

1. Nathan Miller, *The U.S. Navy: A History*, 3rd ed. (Annapolis, Md.: Naval Institute Press, 1977, 1990, 1997), 143–146.

2. Vice Admiral Harry McL. P. Huse, 75th Anniversary, July 25, 1932, John L. Sprague Collection.

3. Jack Sweetman, *The U.S. Naval Academy: An Illustrated History* (Annapolis, Md.: Naval Institute Press, 1979), 83–111.

4. Register of the United States Naval Academy, 1875–1876, 30.

5. Frank Julian Sprague, *Frank J. Sprague: Seventy-fifth Anniversary*, 29–30.

6. The sloop of war *Constellation* was launched in 1854 at the Gosport Navy Yard at Portsmouth, Virginia, and served as flagship of an anti-slavery patrol and during the Civil War. Converted to practice ship service, it served as a Naval Academy practice ship for 22 years, from 1872 to 1893. It was taken out of commission in 1893 but remains today in fully restored condition at Baltimore's Inner Harbor. The historic ship ranks as both the last all-sail ship in the United States Navy and the only remaining Navy Civil War ship afloat.

7. Register of the United States Naval Academy, 1875–1876, 28–29.

8. Register of the United States Naval Academy, 1877–1878, 31–32.

9. Letter from Frank J. Sprague to Mattie H. Munro, Nov. 12, 1876, Frank J. Sprague Papers (Collection number 628), East Carolina Manuscript Collection, Special Collections Department, J. Y. Joyner Library, East Carolina University, Greenville, N.C.

10. Ibid.

11. Park Benjamin, *The United States Naval Academy* (New York: G. P. Putnam's Sons, 1900), 291–294.

12. Rear Admiral James H. Glennon, Sprague 75th Anniversary, June 23, 1932, John L. Sprague Collection. None of the three black midshipmen appointed in 1872–1874 completed the Academy program, and no others were appointed until well into the 20th century. The first black midshipman did not graduate from the Academy until the mid-1900s.

13. Naval Historical Center, available at: http://www.history.navy.mil/photos/pers-us/uspers-b/fw-brtlt.htm (accessed March 17, 2009).

14. Naval Historical Center, available at: http://www.history.navy.mil/photos/pers-us/uspers-g/j-glenon.htm (accessed March 17, 2009).

15. Vice Admiral William Ledyard Rodgers, U.S. Navy (deceased), Navy Office of Administration, May 8, 1944.

16. Naval Historical Center, available at: http://www.history.navy.mil/photos/pers-us/uspers-h/hmp-huse.htm (accessed March 17, 2009).

17. University of Michigan, available at: http://www.engin.umich.edu/newscenter/pubs/engineer/04SS/timeline.html (accessed March 17, 2009).

18. Letter to Frank J. Sprague from Thomas A. Edison, May 26, 1878, Frank Julian Sprague Papers, Mss. & Archives Section, N.Y.P.L.

19. Frank Julian Sprague, "Some Personal Experiences," *Street Railway Journal* 24, no. 15 (Oct. 8, 1904): 566.

20. Naval Historical Center, available at: http://www.history.navy.mil/danfs/r6/richmond-ii.htm.

21. Letter to Secretary of the Navy from Captain Burham, Commanding, Mar. 14, 1880, Frank Julian Sprague Papers, Mss. & Archives Section, N.Y.P.L.

22. Incomplete letter of 1879 from Frank J. Sprague, Frank Julian Sprague Papers, Mss. & Archives Section, N.Y.P.L.

23. Letter to "Miss Frankie" from Frank J. Sprague, Dec. 29, 1879, Frank Julian Sprague Papers, Mss. & Archives Section, N.Y.P.L.

24. Frank Julian Sprague, "Digging in 'The Mines of the Motors'," *Electrical Engineering* 53, no. 5 (May 1934): 695–696.

25. Sprague, "Some Personal Experiences," 566.

26. Letter from Frank J. Sprague to Mattie H. Munro, April 16, 1881, Frank J. Sprague Papers, J. Y. Joyner Library.

27. Letter to William H. Hunt, Secretary of the Navy, from Moses G. Farmer, Sept. 5, 1881, Frank Julian Sprague Papers, Mss. & Archives Section, N.Y.P.L.

28. Frank Julian Sprague, "Digging in 'The Mines of the Motors'," 696.

29. Sprague, "Some Personal Experiences," 567.

30. Ibid. A telpherage, invented by Fleming Jenkin, was a light automatic car suspended and operating from aerial cables, usually electric.

31. Harold C. Passer, "Part VIII, Frank Julian Sprague; Father of Electric Traction, 1857–1934," in William Miller, ed., *Men in Business* (Cambridge, Mass.: Harvard University Press), 218–219.

32. Sprague, "Some Personal Experiences," ibid.

33. Ensign Frank J. Sprague, *Report on the Exhibits at the Crystal Palace Electrical Exhibition, 1882*, General Information Series, No. II: Information From Abroad.

Navy Department, Bureau of Navigation, Office of Naval Intelligence, 1884 (Washington, D.C.: Government Printing Office, 1883).

34. Sprague, "Some Personal Experiences," ibid.

3. Sprague and the New World of Electricity

1. Letter to Secretary of the Navy William E. Chandler from Frank J. Sprague, Mar. 12, 1883, Frank Julian Sprague Papers, Mss. & Archives Section, N.Y.P.L.

2. *The National Cyclopedia of American Biography,* V. 33, 475–476.

3. Frank J. Sprague, "Some Personal Experiences," *Street Railway Journal* 24, no. 15 (Oct. 8, 1904): 567.

4. Paul Israel, *Edison: A Life of Invention* (New York: John Wiley & Sons, 1998), 219, and Ronald W. Clark, *Edison: The Man Who Made the Future* (New York: G. P. Putnam's Sons, 1977), 108–110.

5. Paul Israel, *Edison: A Life of Invention,* 223.

6. Frank J. Sprague, "Some Personal Experiences," 567.

7. Phillip A. Lange, "An Event in Electrical Development," extract from an address delivered to the dinner of the Engineers' Club of Manchester, England, Mar. 15, 1907, Frank Julian Sprague Papers, Mss. & Archives Section, N.Y.P.L.

8. Frank J. Sprague, letter to Thomas A. Edison, on Apr. 24, 1884, Frank Julian Sprague Papers, Mss. & Archives Section, N.Y.P.L.

9. Thomas A. Edison, letter to Frank Julian Sprague, on Apr. 24, 1884, Frank Julian Sprague Papers, Mss. & Archives Section, N.Y.P.L.

10. *The Philadelphia Press,* Sept. 21, 1884.

11. Frank J. Sprague, "Digging in 'The Mines of the Motors'," 698–699.

12. Frank J. Sprague, *Frank J. Sprague: Seventy-fifth Anniversary,* 31.

13. Frank J. Sprague, "The Growth of Electric Railways," an address delivered before the Thirty-Fifth Annual Convention of the American Electric Railway Association at Atlantic City, N. J., Oct. 12, 1916, reproduced in *The Growth of Electric Railways: A Historical Review of the Physical Development of One of the Nation's Greatest Industries,* (New York: American Electric Railway Assn., 1916): 17.

14. Harold C. Passer, *The Electrical Manufacturers, 1875–1900: A Study in Competition, Entrepreneurship, Technical Change, and Economic Growth* (Cambridge, Mass.: Harvard University Press, 1953), 238–239.

15. Notes on the Sprague family from John L. Sprague. Letter from Mary K. Das to Harriet C. Sprague, Oct. 12, 1947.

16. For a more detailed description of the early electric railways, see chapter 4.

17. Frank J. Sprague, "Some Personal Experiences," 568.

18. Ibid.

19. Frank J. Sprague, "Application of Electricity to Propulsion of Elevated Railroads," *The Electrical World* 7, no. 3 (Jan. 16, 1886): 27–28; no. 4 (Jan. 23, 1886): 36; no. 10 (Mar. 6, 1886): 106–107; no. 11 (Mar. 13, 1886): 118–119.

20. Frank J. Sprague, *Electric Traction in Space of Three Dimensions.* Reprinted from *The Journal of the Maryland Academy of Sciences* 2, nos. 3 and 4 (Dec. 1931); III, nos. 1, 2, and 3 (July 1932): 190.

21. Frank J. Sprague, "Some Personal Experience," ibid.

22. Frank J. Sprague, "Some Personal Experience," 569.

23. "The Sprague Electric Railway System," *The Electrical World* 7, no. 13 (Sept. 25, 1886): 152.

4. Triumph at Richmond

1. John H. White, Jr., *Horse Cars: City Transit before the Electrical Age* (Oxford, Ohio: Miami University Libraries, 2006).

2. William D. Middleton, *The Time of the Trolley* (Milwaukee, Wisc.: Kalmbach Publishing Co., 1967), 27, 30, 32–35.

3. William D. Middleton, "A Century of Cable Cars," *American Heritage* 36, no. 3 (Apr./May 1985): 90–101.

4. Verbatim report of the Second Annual Meeting of the American Street Railway Association, held at the Grand Pacific Hotel, Chicago, Illinois, Oct. 9–10, 1883.

5. Middleton, *The Time of the Trolley,* 54–65, covers the pioneer phase of North American street railways. John R. Stevens, ed., *Pioneers of Electric Railroading: Their Story in Words and Pictures.* Headlights, Vol. 51 and 52 (Electric Railroaders' Assn., 1991) provides a detailed account of the pioneer phase of both North American and European street railways.

6. W. Earl Long, *Richmond 1888: Dawn*

of the Electric Street Railway Era (Richmond, Va.: Lady Liberty Press, 1988) 10–13.

7. Frank Julian Sprague, "Lessons of the Richmond Electric Railway," The Engineering Magazine 7, no. 6 (Sept. 1894): 789.

8. Letter to Frank J. Sprague from Oscar T. Crosby, at Sprague's 75th birthday celebration, June 23, 1932, John L. Sprague Collection.

9. The Street Railway Gazette, ca. 1894.

10. Sprague, "Lessons of the Richmond Electric Railway," 799.

11. Letter to Frank J. Sprague from S. W. Huff, May 5, 1932, at Sprague's 75th birthday celebration, John L. Sprague Collection.

12. Sprague, "Lessons of the Richmond Electric Railway," 791, 793.

13. Ibid., 794.

14. Ibid.

15. The Richmond Dispatch, Nov. 8, 1887.

16. Sprague, "Lessons of the Richmond Electric Railway," 798.

17. The Richmond Dispatch, Jan. 10, 1888.

18. Sprague, "Lessons of the Richmond Electric Railway," 798–799.

19. Frank Julian Sprague, "The Growth of Electric Railways." An address to the American Electric Railway Association, published 1916, pp. 26–27.

20. Sprague, "Lessons of the Richmond Electric Railway," 799–800.

21. Letter from The Richmond Union Passenger Railway Co. to E. H. Johnson, Esq., President Sprague Electric Railway and Motor Co. of May 15, 1888, Frank Julian Sprague Papers, Mss. & Archives Section, N.Y.P.L.

22. Sprague, "The Growth of Electric Railways," 27.

23. Frank J. Sprague, "The Electric Railway," The Century Magazine 70 (new series 48), no. 4 (Aug. 1905): 520.

24. Sprague, "The Growth of Electric Railways," 27–28.

25. Frank Julian Sprague, "Birth of the Electric Railway," Transit Journal 78, no. 10 (Sept. 15, 1934): 321.

26. Harold C. Passer, Men in Business, "Part VIII, Frank Julian Sprague; Father of Electric Traction, 1857–1934" (New York: Harper and Row, 1952), 226.

27. William Le Roy Emmet, The Autobiography of an Engineer (Albany, N.Y.: Fort Orange Press, 1931), 73.

28. "Edison System of Electric Railways," Edison General Electric Co., 1891, Frank Julian Sprague Papers, Mss. & Archives Section, N.Y.P.L.

29. Letter from Frank Julian Sprague to the President and Board of Directors of the Edison General Electric Company, of Dec. 2, 1890, Frank Julian Sprague Papers, Mss. & Archives Section, N.Y.P.L.

5. Sprague and the Electric Elevator

1. Jason Goodwin, Otis: Giving Rise to the Modern City (Chicago: Ivan R. Dee, Publisher, 2001), 13–16.

2. Ibid., 7–8.

3. Ibid., 17.

4. Lee E. Gray, From Ascending Rooms to Express Elevators: A History of the Passenger Elevator in the 19th Century (Mobile, Ala.: Elevator World, 2002), 32–39.

5. James D. McCabe, New York by Sunlight and Gaslight (New York: Union Publishing House, 1886), 40.

6. Goodwin, 37–51.

7. Carl W. Condit, American Building Art: The Nineteenth Century (New York: Oxford University Press, 1960), 39–63.

8. Frank J. Sprague, The Journal of the Maryland Academy of Sciences 3, no. 1, 2, and 3, July 1932.

9. John Winthrop Hammond, Men and Volts (New York: J. B. Lippincott Co., 1941), 117–118.

10. Goodwin, 75–77.

11. Gray, 188–190.

12. The Electrical Engineer 17 (Apr. 18, 1894): 351.

13. Frank J. Sprague, The Journal of the Maryland Academy of Science, ibid., 14.

14. Frank J. Sprague, Street Railway Journal 24, no. 15 (Oct. 8, 1904): 571.

15. Frank J. Sprague, Transactions of the American Institute of Electrical Engineers 13 (Jan. 22, 1896): 3–22.

16. Ibid., June 12, 1886, 24.

17. Ibid., Feb. 26, 1886, 37–67.

18. Letter to Frank J. Sprague from Louis K. Comstock, May 5, 1932, at Sprague's 75th anniversary celebration, John L. Sprague Collection.

19. Elevators for All Kinds of Service—Sprague Electric Elevators, Publication no. 1 for Foreign Trade, Sprague Electric Co., Apr. 1899, Frank Julian Sprague Papers, Mss. & Archives Section, N.Y.P.L.

20. Gray, 197–199.

21. Ibid., 199–200.

22. Sir Benjamin Baker was the design engineer for the great Firth of Forth railway bridge in Scotland and a few years later the Aswan Dam in Egypt.

23. *Engineering* (Mar. 3, 1899): 273–276; (Mar. 10, 1899): 304–307.

24. Gray, 198–205.

25. Condit, 9–14.

26. "Higher Car Speeds Proposed for Elevators," *Power* (Jan. 6, 1931).

27. Preamble letter from Frank J. Sprague on the dual elevator system, Dec. 15, 1926, Frank Julian Sprague Papers, Mss. & Archives Section, N.Y.P.L.

28. "Electric Control Permits Two Elevators in Same Shaft," *Electrical World* 97, (Feb. 14, 1931): 312–316. Dr. Lee Gray, "Two Cars in One Shaft," *Elevator World* 54, no. 5 (May 2006): 2–7.

29. James W. Fortune, "Mega High-Rise Elevators," *Elevator World* 43, no. 7 (July 1995): 63–69.

30. Nick Paumgarten, "Up and Then Down," *The New Yorker,* Apr. 21, 2008: 106–115.

6. Frank Sprague and the Multiple Unit Train

1. Frank J. Sprague, "Some Personal Experiences," *Street Railway Journal* 24, no. 15 (Oct. 8, 1904): 571.

2. Ibid.

3. Frank J. Sprague, "Considerations Which Should Govern the Selection of a Rapid Transit System," *Transactions of the American Institute of Electrical Engineers* 8 (1891): 331–350.

4. *New-York Daily Tribune,* June 15, 1891, "Mr. Gould on Rapid Transit."

5. Frank J. Sprague, "Some Personal Experiences," 572–573.

6. Frank J. Sprague, letter to George Gould, Russell Sage, and F. M. Gallaway, Special Committee, Manhattan Elevated R.R., Feb. 15, 1897, Frank Julian Sprague Papers, Mss. & Archives Section, N.Y.P.L.

7. Frank Julian Sprague, letter to George Gould, Mar. 11, 1897, Frank Julian Sprague Papers, Mss. & Archives Section, N.Y.P.L.

8. Frank J. Sprague, "Some Personal Experiences," 573.

9. In Chicago the elevated railway was almost always referred to as the "L."

10. William D. Middleton, *Metropolitan Railways: Rapid Transit in America* (Bloomington: Indiana University Press, 2003), 35–36.

11. Bruce G. Moffat, *The "L": The Development of Chicago's Rapid Transit System, 1888–1932* (Chicago: Central Electric Railfans' Assn., 1995), 14–19.

12. Frank J. Sprague, review report to Leslie Carter, President, South Side Elevated Railway Co., dated Apr. 7, 1897, Frank Julian Sprague Papers, Mss. & Archives Section, N.Y.P.L.

13. Ibid.

14. Frank J. Sprague, "Digging in 'The Mines of the Motors'," *Electrical Engineering* 53, no. 5 (May 1934): 704–705.

15. Moffat, 36–37.

16. C. R. McKay, letter to Frank J. Sprague, dated May 5, 1897, Frank Julian Sprague Papers, Mss. & Archives Section, N.Y.P.L.

17. Letter from Willits H. Sawyer to Frank J. Sprague, on July 25, 1932, John L. Sprague Collection.

18. Letter from Samuel B. Stewart to Frank L. Sprague, on June 7, 1932, John L. Sprague Collection.

19. "The New Sprague Multiple Unit System of Car Traction," *The Electrical Engineer* 24, no. 482 (July 29, 1897): 93–94.

20. "The New Sprague Company," *The Electrical Engineer* 24, no. 493 (Oct. 14, 1897): 355.

21. Letter from Fred W. Butt to Frank J. Sprague, on June 14, 1932, John L. Sprague Collection.

22. Moffat, 38.

23. Frank J. Sprague, "The Multiple Unit System of Electric Railways," *Transactions of the American Institute of Electrical Engineers* 16 (1900): 193–249.

24. Ibid., 202–203.

25. Ibid., 206.

26. Ibid., 209.

27. Ibid.

28. Frank J. Sprague, "Some Personal Experiences," 574.

29. C. R. McKay, letter to Frank J. Sprague.

30. "Sprague Multiple Unit Control and the Adoption of Electricity on the Brooklyn Elevated System," *The Electrical Engineer* 25, no. 515 (Mar. 17, 1898): 301–302.

31. "Electrical Equipment of the Brook-

lyn Elevated Railroad," *The Electrical Engineer* 25, no. 513 (Mar. 3, 1898): 512–513.

32. Middleton, ibid., 40.

33. Ibid., 42–44.

34. Ibid., 40–42.

35. "Apropos the New York Subway and Other Rapid Transit Problems and a Few Facts Ré Sprague in Connection Therewith." Manuscript, ca. 1916. Frank Julian Sprague Papers, Shore Line Trolley Museum.

36. Harold C. Passer, *The Electrical Manufacturers 1875–1900; A Study in Competition, Entrepreneurship, Technical Change, and Economic Growth* (Cambridge, Mass.: Harvard University Press, 1953), 274–275.

37. "Absorption of Sprague Electric by General Electric," *Electrical World and Engineer* 29, no. 22 (May 31, 1902): 971–972.

7. Electrifying the Main Line Railroads

1. Frank J. Sprague, "Coming Development of Electric Railways, I-IV," *The Electrical Engineer* 13, no. 215 (June 15, 1892): 611–612; no. 216 (June 22, 1892): 629–630; no. 217 (June 29, 1892): 661–662; and 14, no. 218 (July 6, 1892): 12.

2. Alexandra Villard De Borchgrave and John Cullen, *Villard: The Life and Times of an American Titan* (New York: Doubleday, 2001): 320–324.

3. Although the windmill-powered generators proposed by Edison were never built, the idea of wind-powered electric trains is still around. In 2008 the Burlington Northern Sante Fe Railroad began studying the feasibility of using its right-of-way for high-tension power lines that would transmit power from huge wind-powered generators on the high winds of the Great Plains east of the Rocky Mountains to service the high power demands in California. The transmission lines along the BNSF could then also deliver power for electric trains.

4. John R. Stevens, ed., *Pioneers of Electric Railroading: Their Story in Words and Pictures,* 37–42.

5. *The National Cyclopedia of American Biography,* 145–146.

6. Frank J. Sprague, "Coming Development of Electric Railways—IV," *The Electrical Engineer* 14, no. 218 (July 6, 1892): 12.

7. "The 1,000 H.P. Sprague Electric Locomotive," *The Electrical Engineer* 16, no. 285 (Oct. 18, 1893): 339–341.

8. Notes on the Sprague family from John L. Sprague.

9. Letter to Edwin Aday from Frank J. Sprague, Feb. 11, 1911, Frank Julian Sprague Papers, Mss. & Archives Section, N.Y.P.L.

10. Letter to Frank J. Sprague from Albert Bigelow Paine, Apr. 24, 1932, at Sprague's 75th birthday celebration, John L. Sprague Collection.

11. Notes on the Sprague family from John L. Sprague.

12. William D. Middleton, *Grand Central: The World's Greatest Railroad Terminal* (San Marino, Calif.: Golden West Books, 1977), 20–51.

13. Letter to Frank J. Sprague from William J. Wilgus, May 5, 1932, at Sprague's 75th birthday celebration, John L. Sprague Collection.

14. Letter to William J. Wilgus from Frank J. Sprague, Feb. 8, 1902, Frank Julian Sprague Papers, Mss. & Archives Section, N.Y.P.L.

15. Middleton, *Grand Central,* 52–61.

16. Letter to B. J. Arnold, Frank J. Sprague, and George Gibbs from William J. Wilgus, Dec. 15, 1902, Frank Julian Sprague Papers, Mss. & Archives Section, N.Y.P.L.

17. Statement by Frank Julian Sprague as part of the minutes of the Forty-First Regular Meeting of the Electric Traction Committee, Nov. 3, 1903, Frank Julian Sprague Papers, Mss. & Archives Section, N.Y.P.L., 1–4.

18. Ibid., 4–10.

19. Letter to Frank J. Sprague from William J. Wilgus, Jan. 15, 1910, Frank Julian Sprague Papers, Mss. & Archives Section, N.Y.P.L.

20. William D. Middleton, *When the Steam Railroads Electrified,* 2nd ed. (Bloomington: Indiana University Press, 2001), 40–44.

21. *The New York Central Electrification,* General Electric Company, GEA-902 (Schenectady, N.Y.: General Electric Co., Jan. 1929).

22. Frank J. Sprague, "The Transmission of Power by Electricity," *Journal of the Franklin Institute* 127, nos. 3 and 4 (Mar. 1889 and Apr. 1889): 161–176, 254–264.

23. Letter to Frank J. Sprague from Guido Panteleoni, June 30, 1932, at Sprague's 75th birthday celebration, John L. Sprague Collection.

24. Frank J. Sprague, "Coming Development of Electric Railways," *The Electrical Engineer* 13, no. 215, no. 216, no. 217; 14, no. 218 (June 15, 1892; June 22, 1892; June 29, 1892; July 6, 1892): 611–612, 629–630, 661–662, 712.

25. Frank J. Sprague, "Some Facts and Problems Bearing on Electric Trunk Line Opera-

tion," *Transactions of the American Institute of Electrical Engineers* 26, Part 1 (1908): 681–812.

26. *The New York Central Electrification*, 3–11.

27. Frank J. Sprague, "Electric Traction in Space of Three Dimensions," *The Journal of the Maryland Academy of Sciences* 2, nos. 3 and 4 (Dec. 1931) and 3, nos. 1, 2, and 3 (July 1932): 24–34.

28. Maury Klein, *The Life and Legend of E. H. Harriman* (Chapel Hill: University of North Carolina Press, 2000), 254–265.

29. Robert S. Ford, *Red Trains in the East Bay*, Interurbans Special 65 (Glendale, Calif.: 1977): 100–129.

30. "Harriman and the Southern Pacific," *Street Railway Journal* 29, no. 4 (Jan. 26, 1907): 165.

31. "Southern Pacific Company," *Street Railway Journal* 30, no. 19 (Dec. 14, 1907): 1154.

32. "Electrification on the Southern Pacific," *The Railroad Gazette* 43, no. 10 (Sept. 6, 1907): 249–250.

33. Ibid.

34. "Engineering Board to Report on Electrification for Southern Pacific," *The Railway Age* 44, no. 10 (Sept. 6, 1907): 329.

35. "General News Section," *Railroad Age Gazette* 47, no. 3 (July 16, 1909): 113.

36. "General News Section," *Railroad Age Gazette* 47, no. 4 (July 23, 1909): 159.

37. "Comparative Tests of Consolidation and Mallet Locomotives on Southern Pacific," *Railway Age Gazette* 48, no. 2 (Jan. 14, 1910): 91–94.

38. "General News Section," *Railway Age Gazette* 48, no. 2 (Jan. 14, 1910): 101.

39. Letter to William Hood from Frank J. Sprague, Feb. 1, 1910, Frank Julian Sprague Papers, Mss. & Archives Section, N.Y.P.L.

40. Letter to Frank J. Sprague from William Hood, Feb. 8, 1910, Frank Julian Sprague Papers, Mss. & Archives Section, N.Y.P.L.

41. Letter to Julius Kruttschnitt from Frank J. Sprague, July 11, 1910, Frank Julian Sprague Papers, Mss. & Archives Section, N.Y.P.L.

42. Letter to Frank J. Sprague from Julius Kruttschnitt. Aug. 5, 1910, Frank Julian Sprague Papers, Mss. & Archives Section, N.Y.P.L.

43. Clifton Hood, *722 Miles: The Building of the Subways and How They Transformed New York* (New York: Simon & Schuster, 1993), 136–150.

44. "Proposal for Equipment and Operation of an Independent City-Built Rapid Transit System," from Frank J. Sprague and Oscar T. Crosby, Jan. 25, 1911, from Frank Julian Sprague Papers, Shore Line Trolley Museum.

8. The Naval Consulting Board and the Great War

1. Letter from Secretary of the Navy J. Daniels to T. A. Edison, July 7, 1915, reprinted in: Lloyd N. Scott, *Naval Consulting Board of the United States* (Washington, D.C.: Government Printing Office, 1920), 286–288.

2. Ibid.

3. Letter from Frank J. Sprague to J. C. Parker, July 31, 1915, Frank Julian Sprague Papers, Mss. & Archives Section, N.Y.P.L.

4. Scott, 1920, 11.

5. Letter from Frank J. Sprague to J. C. Parker, July 31, 1915, Frank Julian Sprague Papers, Mss. & Archives Section, N.Y.P.L.

6. Scott, 1920, 11–13.

7. *New York Times*, Dec. 23, 1915.

8. *Columbus Dispatch*, Nov. 13, 1915.

9. *Schenectady Union Star*, Dec. 9, 1915; *Pittsburgh Press*, Dec. 16, 1915.

10. *Brooklyn Citizen*, Dec. 23, 1915; *Baltimore Evening Sun*, Dec. 23, 1915.

11. *New York World*, Dec. 24, 1915; *New York Press*, Dec. 24, 1915.

12. Letter from Frank J. Sprague to T. Robbins, Apr. 25, 1916, Frank Julian Sprague Papers, Mss. & Archives Section, N.Y.P.L.

13. "Naval Consulting Board Personnel," *Engineering Magazine* 50, no. 2 (Nov. 1915): 199–221.

14. Letter from J. R. Dunlap to J. Daniels, Oct. 4, 1915, and Oct. 11, 1915, Frank Julian Sprague Papers, Mss. & Archives Section, N.Y.P.L.

15. Letter from J. R. Dunlap to Frank J. Sprague, Oct. 5, 1915, Frank Julian Sprague Papers, Mss. & Archives Section, N.Y.P.L.

16. Scott, 1920, 13.

17. Scott, 1920, 14–15.

18. Various correspondence from T. Robbins to the members of the Board, Frank Julian Sprague Papers, Mss. & Archives Section, N.Y.P.L.

19. Letter from Frank J. Sprague to T. Robbins, Nov. 30, 1915, Frank Julian Sprague Papers, Mss. & Archives Section, N.Y.P.L.

20. Letter from T. Robins to J. Daniels, Oct. 21, 1915, Frank Julian Sprague Papers, Mss. & Archives Section, N.Y.P.L.

21. Letter from Frank J. Sprague to T. Robbins, Dec. 16, 1915, Frank Julian Sprague Papers, Mss. & Archives Section, N.Y.P.L.

22. Oath of Office, Frank J. Sprague, Naval Consulting Board. Sept. 19, 1916, Frank Julian Sprague Papers, Mss. & Archives Section, N.Y.P.L.

23. Letter from Frank J. Sprague to L. Baekeland, Dec. 16, 1915, Frank Julian Sprague Papers, Mss. & Archives Section, N.Y.P.L.

24. Letter from L. Baekeland to Frank J. Sprague, Dec. 20, 1915, Frank Julian Sprague Papers, Mss. & Archives Section, N.Y.P.L.

25. Pamphlet prepared by the Naval Consulting Board presenting the views of the various Board members on their thoughts about the Board and its future, Apr. 1919, Frank Julian Sprague Papers, Mss. & Archives Section, N.Y.P.L.

26. Naval Advisory Board, Dec. 1915, Frank Julian Sprague Papers, Mss. & Archives Section, N.Y.P.L.

27. Scott, 1920, 29.

28. Ibid., 30–35.

29. Ibid., 36–37.

30. Ibid., 38–39.

31. *New York Times,* Jan. 13, 1916.

32. *Philadelphia Evening Telegram,* Jan. 12, 1916.

33. *New York Press,* Feb. 6, 1916.

34. *Albany Argus,* Feb. 6, 1916.

35. *The World,* Feb. 6, 1916.

36. *Providence Journal,* Feb. 6, 1916.

37. Scott, 1920.

38. Ibid., 56–57.

39. Ibid., 58–60.

40. Letter from H. Maxim to S. Miller, Oct. 27, 1917, Frank Julian Sprague Papers, Mss. & Archives Section, N.Y.P.L.

41. Scott, 1920, 67–83.

42. Ibid., 84–108.

43. Ibid., 123–124.

44. Letter from Henry T. Weed, Principal of the Brooklyn Evening Technical and Trade School for Men and Women to T. Robbins, May 17, 1917, Frank Julian Sprague Papers, Mss. & Archives Section, N.Y.P.L.

45. Scott, 1920, 125.

46. Letter from Sprague to W. L. Saunders, Jan. 2, 1920, Frank Julian Sprague Papers, Mss. & Archives Section, N.Y.P.L.

47. Letter to Secretary Robbins, July 18, 1917, Frank Julian Sprague Papers, Mss. & Archives Section, N.Y.P.L.

48. Letter to Secretary Robbins, Aug. 24, 1917, in response to a request of Aug. 23 from Secretary Robbins that members of the Board state their present or contemplated dealings with the government. Sprague also outlines a series of inventions that he is actively engaged in, and states that these will only be acted on by the Board in accordance with the protocols in place for all ideas submitted to the Board. Frank Julian Sprague Papers, Mss. & Archives Section, N.Y.P.L.

49. Letter to T. Robins from Frank J. Sprague, Mar. 27, 1916, Frank Julian Sprague Papers, Mss. & Archives Section, N.Y.P.L.

50. Correspondence from A. W. Burns, Frank J. Sprague, and H. Maxim, 1918, Frank Julian Sprague Papers, Mss. & Archives Section, N.Y.P.L.

51. Ian R. McNab, "Early Electric Gun Research," *ISEE Transactions on Magnetics* 35, no. 1 (1999),: 250–261.

52. Ibid.

53. Correspondence between B. G. Lamme and Frank J. Sprague, Apr. and May 1917, Frank Julian Sprague Papers, Mss. & Archives Section, N.Y.P.L.

54. Letter from Secretary Robbins to Frank J. Sprague, May 23, 1917, Frank Julian Sprague Papers, Mss. & Archives Section, N.Y.P.L.

55. Letter to President Wilson, July 20, 1917, Frank Julian Sprague Papers, Mss. & Archives Section, N.Y.P.L.

56. Various correspondence, Frank Julian Sprague Papers, Mss. & Archives Section, N.Y.P.L.

57. "Proposal for the Defeat of Submarines: Based upon an Extension of Naval Strategy, requiring a New Type and Size of Boat, and including a new war Weapon," July 7, 1917, Frank Julian Sprague Papers, Mss. & Archives Section, N.Y.P.L.

58. Letter from Frank J. Sprague to B. G. Lamme, Sept. 12, 1917, Frank Julian Sprague Papers, Mss. & Archives Section, N.Y.P.L.

59. Various correspondence, Frank Julian Sprague Papers, Mss. & Archives Section, N.Y.P.L.

60. *New York Times,* Dec. 28 and 29, 1915.

61. Letter from Frank J. Sprague to L. N.

Scott, Aug. 28, 1919, Frank Julian Sprague Papers, Mss. & Archives Section, N.Y.P.L.

62. Correspondence between Frank J. Sprague and Lindon Bates (Submarine Defense Association) between Sept. 1917 and July 1918, Frank Julian Sprague Papers, Mss. & Archives Section, N.Y.P.L.

63. Letter from Frank J. Sprague to B. G. Lamme, June 2, 1917, Frank Julian Sprague Papers, Mss. & Archives Section, N.Y.P.L.

64. Correspondence between Frank J. Sprague and Lindon Bates (Submarine Defense Association) between Sept. 1917 and July 1918, Frank Julian Sprague Papers, Mss. & Archives Section, N.Y.P.L.

65. Letter from Frank J. Sprague to Rear-Admiral Ralph Earle, Oct. 19, 1918, Frank Julian Sprague Papers, Mss. & Archives Section, N.Y.P.L.

66. Report from the Special Board on Naval Ordnance (R. R. Ingersoll) to the Chief of Bureau (Rear-Admiral Ralph Earle), Nov. 20, 1918, Frank Julian Sprague Papers, Mss. & Archives Section, N.Y.P.L.

67. Letter from Frank J. Sprague to Rear-Admiral Ralph Earle, Oct. 19, 1918, Frank Julian Sprague Papers, Mss. & Archives Section, N.Y.P.L.

68. Memorandum from the Naval Bureau of Ordnance, Oct. 4, 1915, Frank Julian Sprague Papers, Mss. & Archives Section, N.Y.P.L.

69. Memorandum from the Naval Bureau of Construction and Repair, Oct. 4, 1915, Frank Julian Sprague Papers, Mss. & Archives Section, N.Y.P.L.

70. Undated Memorandum from the Naval Bureau of Steam Engineering, Frank Julian Sprague Papers, Mss. & Archives Section, N.Y.P.L.

71. Letter from L. H. Baekeland to A. M. Hunt, Dec. 30, 1915, Frank Julian Sprague Papers, Mss. & Archives Section, N.Y.P.L.

72. Letter from F. J. Sprague to T. Robbins, Dec. 28, 1915, Frank Julian Sprague Papers, Mss. & Archives Section, N.Y.P.L.

73. Letter from Frank J. Sprague to T. Robbins, Mar. 27, 1916, Frank Julian Sprague Papers, Mss. & Archives Section, N.Y.P.L.

74. Correspondence between L. H. Baekeland and Frank J. Sprague, Mar. 13–15, 1916, Frank Julian Sprague Papers, Mss. & Archives Section, N.Y.P.L.

75. Various correspondence, Dec. 6 and 7, 1916, Frank Julian Sprague Papers, Mss. & Archives Section, N.Y.P.L.

76. Letters from W. F. M. Goss to Frank J. Sprague, Oct.–Nov. 1916, Frank Julian Sprague Papers, Mss. & Archives Section, N.Y.P.L.

77. Report of the Committee on Sites, Dec. 14, 1916, Frank Julian Sprague Papers, Mss. & Archives Section, N.Y.P.L.

78. Report to T. Robbins by T. A. Edison on the Laboratory Site, Oct. 7, 1916, Frank Julian Sprague Papers, Mss. & Archives Section, N.Y.P.L.

79. Letters from L. H. Baekeland to Frank J. Sprague, Dec. 5 and 6, 1916, Frank Julian Sprague Papers, Mss. & Archives Section, N.Y.P.L.

80. Letter to Frank J. Sprague from T. Robbins, Dec. 12, 1916, Frank Julian Sprague Papers, Mss. & Archives Section, N.Y.P.L.

81. Letter to T. A. Edison from Frank J. Sprague, Dec. 13, 1916, Frank Julian Sprague Papers, Mss. & Archives Section, N.Y.P.L.

82. Letter to W. R. Whitney from Frank J. Sprague, Dec. 14, 1916, Frank Julian Sprague Papers, Mss. & Archives Section, N.Y.P.L.

83. Frank J. Sprague, Suggested Report, Oct. 17, 1916, Frank Julian Sprague Papers, Mss. & Archives Section, N.Y.P.L.

84. Letter from T. Robbins to the members of the Committee on Sites, Nov. 15, 1916, Frank Julian Sprague Papers, Mss. & Archives Section, N.Y.P.L.

85. Frank J. Sprague, Report of the Committee on Sites, Dec. 14, 1916, Frank Julian Sprague Papers, Mss. & Archives Section, N.Y.P.L.

86. Ibid.

87. Letter to J. W. Weeks from Frank J. Sprague, Mar. 15, 1918, Frank Julian Sprague Papers, Mss. & Archives Section, N.Y.P.L.

88. Report to the Laboratory Committee, S. H. Condict, June 3, 1918, Frank Julian Sprague Papers, Mss. & Archives Section, N.Y.P.L.

89. Letter to T. A. Edison from Frank J. Sprague, Jan. 30, 1918, Frank Julian Sprague Papers, Mss. & Archives Section, N.Y.P.L.

90. Letter to W. L. Saunders from T. Robbins, Dec. 30, 1918, Frank Julian Sprague Papers, Mss. & Archives Section, N.Y.P.L.

91. Letter to J. Daniels from Frank J. Sprague, Jan. 6, 1920, Frank Julian Sprague Papers, Mss. & Archives Section, N.Y.P.L.

92. Letter from W. S. Smith to Frank J.

Sprague, Jan. 7, 1920, Frank Julian Sprague Papers, Mss. & Archives Section, N.Y.P.L.

93. Letter to Frank J. Sprague from J. Daniels, Jan. 19, 1920, Frank Julian Sprague Papers, Mss. & Archives Section, N.Y.P.L.

94. Letter to A. G. Webster from Frank J. Sprague, Mar. 6, 1920, Frank Julian Sprague Papers, Mss. & Archives Section, N.Y.P.L.

95. Letter to Frank J. Sprague from T. Robbins, Apr. 6, 1917, Frank Julian Sprague Papers, Mss. & Archives Section, N.Y.P.L.

96. Correspondence between Frank J. Sprague, J. Daniels, and various board members, 1917, Frank Julian Sprague Papers, Mss. & Archives Section, N.Y.P.L.

97. Letter to L. H. Baekeland from Frank J. Sprague, Dec. 16, 1915, Frank Julian Sprague Papers, Mss. & Archives Section, N.Y.P.L.

98. Letter to L. H. Baekeland from T. A. Edison, Nov. 17, 1915, Frank Julian Sprague Papers, Mss. & Archives Section, N.Y.P.L.

99. Letter from Frank J. Sprague to L. N. Scott, Aug. 28, 1919, Frank Julian Sprague Papers, Mss. & Archives Section, N.Y.P.L.

100. Letter from Frank J. Sprague to L. N. Scott, Dec. 11, 1920, Frank Julian Sprague Papers, Mss. & Archives Section, N.Y.P.L.

101. Letter from Frank J. Sprague to J. Daniels, Jan. 16, 1920, Frank Julian Sprague Papers, Mss. & Archives Section, N.Y.P.L.

9. Sprague and Railroad Safety

1. William D. Middleton, "Technology and Operating Practice in the Twentieth Century," *Encyclopedia of North American Railroads* (Bloomington: Indiana University Press, 2007), 53.

2. Mark Aldrich, *Death Rode the Rails: American Railroad Accidents and Safety, 1828–1965* (Baltimore, Md.: Johns Hopkins University Press, 2006), 120.

3. Mark Aldrich, "Accidents," *Encyclopedia of North American Railroads*, 89, table 1.

4. ICC report on track signaling mileage, Jan. 1, 1908, from *The Railroad Gazette* 44, no. 16 (Apr. 17, 1908): 543–548.

5. "The Block Signal and Train-Control Board," *The Railroad Gazette* 44, no. 2 (Jan. 10, 1908): 52–53.

6. J. B. Latimer, "Train Operation and Automatic Train Control," *Railway Age* 70, no. 16 (Apr. 22, 1921): 977–979.

7. G. E. Ellis et al., "Automatic Train Control," *Proceedings: The Journal of the Pacific Railway Club* 11, no. 4 (July 1927): 5–6.

8. "Signals and Automatic Train Stops in the Hudson & Manhattan Tunnel," *The Railroad Gazette* 44, no. 10 (Mar. 10, 1908): 317.

9. "Automatic Train Stops in New York," *Railway Age* 72, no. 6 (Feb. 11, 1922): 395.

10. Letter to Frank J. Sprague from Lawrence Langner, undated, at Sprague's 75th birthday celebration, John L. Sprague Collection.

11. Letter to Hon. Henry C. Hall, Interstate Commerce Commission, from O. B. Wilcox, Sprague Safety Control and Signal Corp., June 23, 1915, Frank Julian Sprague Papers, Mss. & Archives Section, N.Y.P.L.

12. Ibid.

13. Sprague Auxiliary Train Control, Bulletin No. 3 (Mar. 15, 1923): 3–9; "Great Northern Train Control," *Railway Signaling* 18, no. 1 (Jan. 1925): 7–11; Frank J. Sprague, "Automatic Train Control," *Journal of The Franklin Institute* 194, no. 2 (Aug. 1922): 144–155.

14. "The Sprague System of Automatic Train Control," *Railway Review* (May 27, 1922): 747–756.

15. Letter to Hon. William McAdoo, Director General of Railways, from Frank J. Sprague, President of Sprague Safety Control & Signal Corp., July 11, 1918, Frank Julian Sprague Papers, Mss. & Archives Section, N.Y.P.L.

16. Letter to Senator Albert B. Cummins, Chairman, Senate Committee on Interstate Commerce, from Frank J. Sprague, Oct. 29, 1919, Frank Julian Sprague Papers, Mss. & Archives Section, N.Y.P.L.

17. John J. Esch was a commissioner of the Interstate Commerce Commission and a principal author of Section 26 of the Transportation Act of 1920.

18. Frank J. Sprague, "The Need for Automatic Train Control," *Railway Age* 68, no. 6 (Feb. 6, 1920): 401.

19. Ibid., 401–402.

20. Letters to W. P. Borland, Chief Bureau of Safety, Interstate Commerce Commission, from Frank J. Sprague, President, Sprague Safety Control & Signaling Corp., of June 8, and July 1, 1920, Frank Julian Sprague Papers, Mss. & Archives Section, N.Y.P.L.

21. Letter to A. L. Humphrey, President, Westinghouse Air Brake Co., from Frank J. Sprague, President, Sprague

Safety Control & Signaling Corp., of Jan. 8, 1920, Frank Julian Sprague Papers, Mss & Archives Section, N.Y.P.L.

22. "New York Commission on Automatic Train Control," *Railway Age* 70, no. 9 (Mar. 4, 1921): 497–498.

23. "The Sprague System of Auxiliary Train Control," *Railway Age* 72, no. 16 (Apr. 22, 1922): 963–967.

24. "Report on Sprague Automatic Train Control," *Railway Age* 75, no. 9 (Sept. 1, 1923): 399–400.

25. Letter to C. O. Bradshaw, Asst. Gen. Manager, Chicago, Milwaukee & St. Paul Railway, from John P. Kelly, Senior Mechanical Engineer, Interstate Commerce Commission, on Dec. 24, 1923, Frank Julian Sprague Papers, Mss. & Archives Section, N.Y.P.L.

26. *The Christian Science Monitor,* (Apr. 19, 1924).

27. "Calls for Greater Safety on Railways," *New York Times* (Jan. 12, 1922).

28. Ellis, "Automatic Train Control," 19–27.

29. "Four Objections to Installing Automatic Stops," *Railway Age* 72, no. 12 (Mar. 25, 1922): 783–784.

30. "The New Train Control Order," *Railway Age* 76, no. 5 (Feb. 2, 1924): 314.

31. "What the Roads Have Done with Train Control," *Railway Age* 76, no. 12 (Mar. 12, 1924): 637–639.

32. "No Train Control Installations Ordered," *Railway Age* 85, no. 23 (Nov. 8, 1928): 1139–1142.

33. "Rapid Progress in Automatic Train Control in 1926," *Railway Age* 82, no. l (Jan. 1, 1927): 130–133.

34. "Block Signal Mileage, 1928," *Railway Age* 85, no. 15 (Oct. 13, 1928): 706.

35. G. E. Ellis, "The Status of Train Control," *Railway Age* 83, no. 2 (July 9, 1927): 69–70.

36. "The Train Control Situation," *Railway Age* 89, no. 19 (Nov. 8, 1930): 969.

37. "Should Railroads Advertise Automatic Train Control?" *Railway Age* 83, no. 3 (July 16, 1927): 88.

38. "Rapid Progress in Automatic Train Control in 1926," *Railway Age,* 130–133.

39. Frank J. Sprague, "Digging in 'The Mines of the Motors'," *Electrical Engineering* 53, no. 5 (May 1934): 695–706.

40. H.R. 2095 Fact Sheet, Association of American Railroads, Oct. 2008.

10. A Diverse Inventor

1. Frank Julian Sprague Papers, Mss. & Archives Section, N.Y.P.L.

2. "Some inventions, all save one disclosed in old midshipmen's note book by Frank J. Sprague in the period from 1877 to 1880 (Age 20 to 23)." Frank Julian Sprague Papers, Mss. & Archives Section, N.Y.P.L.

3. "Re; Sprague inventions and developments." Frank Julian Sprague Papers, Mss. & Archives Section, N.Y.P.L.

4. Experimental Journal of Harold B. Cockerline, Frank Julian Sprague Papers, Mss. & Archives Section, N.Y.P.L.

5. Shop Logs, Frank Julian Sprague Papers, Mss. & Archives Section, N.Y.P.L.

6. Frank J. Sprague, "Notes on Automatic Reversing Razor Strop, Oct. 18, 1911" and "Memoranda re Roller Case for Razor Strop," Frank Julian Sprague Papers, Mss. & Archives Section, N.Y.P.L.

7. Frank Julian Sprague Papers, Mss. & Archives Section, N.Y.P.L.

8. Ibid.

9. Ibid.

10. Letter of Recommendation for Harold B. Cockerline, Oct. 15, 1917, Frank Julian Sprague Papers, Mss. & Archives Section, N.Y.P.L.

11. John R. Stevens, ed., *Pioneers of Electric Railroading: Their Story in Words and Pictures,* 45–46.

12. Drawing by Frank Julian Sprague dated Nov. 20, 1888, Frank Julian Sprague Papers, Mss. & Archives Section, N.Y.P.L.

13. "Re; Sprague inventions and developments." Frank Julian Sprague Papers, Mss. & Archives Section, N.Y.P.L

14. "Apropos the New York Subway and other electric railway problems, and a few facts RE Sprague in connection therewith." Undated. Frank Julian Sprague Papers, Seashore Trolley Museum.

15. Ibid., Apr.–May 1887.

16. Ibid., June 13, 1891.

17. Ibid., Feb. 8, 1896.

18. Ibid., Jan. 18, Feb. 1, 1902.

19. Ibid., Mar. 1902.

20. Ibid. Résumé.

21. Newspaper clipping—no date or attribution. "Dead Engineer Holds Throttle: When Big 4 Passenger Misses Accustomed Stop Fireman is Led to Investigate." Shop Records

Book No. 11: 1914–1915, Frank Julian Sprague Papers, Mss. & Archives Section, N.Y.P.L.

22. Frank Julian Sprague Papers, Mss. & Archives Section, N.Y.P.L.

23. Letter to Frank J. Sprague from U.S. Patent Office dated Feb. 27, 1915, Frank Julian Sprague Papers, Mss. & Archives Section, N.Y.P.L.

24. Letter to U.S. Patent Office from Dorsey & Cole (Frank J. Sprague patent attorneys), June 25, 1915, Frank Julian Sprague Papers, Mss. & Archives Section, N.Y.P.L.

25. Letter to Frank J. Sprague from U.S. Patent Office, July 30, 1915, Frank Julian Sprague Papers, Mss. & Archives Section, N.Y.P.L.

26. Various correspondence between Frank J. Sprague, Dorsey & Cole, and the U.S. Patent Office, Frank Julian Sprague Papers, Mss. & Archives Section, N.Y.P.L.

27. Notebook, engineering notes and data, Desmond Sprague, Chief Engineer, Sprague Safety Control and Signal Co., Frank Julian Sprague Papers, Mss. & Archives Section, N.Y.P.L.

28. Letter from U.S. Patent Office to Dorsey & Cole (Frank J. Sprague patent attorneys) dated May 29, 1929, Frank Julian Sprague Papers, Mss. & Archives Section, N.Y.P.L.

11. An Inventor and Engineer to the End

1. Letter to L. C. Sprague from Frank J. Sprague, Dec. 2, 1924, Frank Julian Sprague Papers, Mss. & Archives Section, N.Y.P.L.

2. Letter to Frank J. Sprague from L. C. Sprague, July 29, 1931, Frank Julian Sprague Papers, Mss. & Archives Section, N.Y.P.L.

3. Letter to Mary Das from Frank J. Sprague, Dec. 15, 1933, John L. Sprague Collection.

4. Letter to Frank J. Sprague from William J. Wilgus, Feb. 7, 1934, Frank Julian Sprague Papers, Mss. & Archives Sen, N.Y.P.L.

5. Robert D. Potter, "Father of the Rapid Transit," *New York Herald Tribune, Magazine,* July 24, 1932.

6. *Frank J. Sprague: Seventy-fifth Anniversary.*

7. Letter to Mary Das from Frank J. Sprague, Dec. 15, 1933, John L. Sprague Collection.

8. Donald N. Dewees, "The Decline of the American Street Railways," *Traffic Quarterly* (Oct. 1970): 563–581.

9. William D. Middleton, *The Time of the Trolley,* 126–129.

10. Western Union telegram to Frank

J. Sprague from F. H. Shepard, Sept. 12, 1934, Frank Julian Sprague Papers, Mss. & Archives Section, N.Y.P.L.

11. "New Principles of Car Design A Great Forward Step," *Transit Journal* 78, no. 11 (Oct. 1934): 381–382.

12. Stephen P. Carlson and Fred W. Schneider III, *PCC: The Car that Fought Back.* Interurbans Special 64 (Glendale, Calif.: Interurban Press, 1980).

13. *The New York Times* and *New York Herald Tribune* obituaries, Oct. 26, 1924.

14. Harriet Sprague, *Frank J. Sprague and the Edison Myth.*

12. Epilogue

1. *The Philadelphia Press,* Sept. 21, 1884.

2. Frank J. Sprague, letter to the President and Board of Directors of the Edison General Electric Company, on Dec. 2, 1890, Frank Julian Sprague Papers, Mss. & Archives Section, N.Y.P.L.

3. Letter to Gerard Swope from Frank J. Sprague, Oct. 27, 1931, Frank Julian Sprague Papers, Mss. & Archives Section, N.Y.P.L.

4. Letter to Frank J. Sprague from Gerard Swope, Dec. 30, 1931, Frank Julian Sprague Papers, Mss. & Archives Section, N.Y.P.L.

5. *The Electrical World,* Aug. 9, 1884.

6. Roger Burlingame, *Engineers of Democracy: Inventions and Society in Mature America* (New York: Charles Scribner's Sons, 1940), 237.

7. Edward Marshall, "Edison Sees Higher Fares as Only Solution of Street Car Problem," *The Sun,* Aug. 10, 1919.

8. Frank J. Sprague, "Inventors of the Electric Railway," *The Sun,* Aug. 27, 1919.

9. Thomas A. Edison, "The Great Men of the Electrical Age," *The Sun,* Sept. 4, 1919.

10. Frank J. Sprague, "The Great Men of the Electrical Age," *The Sun,* Sept. 18, 1919.

11. James C. Young, "The Magic Edison Made for the World," *The New York Times Magazine,* Aug. 12, 1928.

12. Frank J. Sprague, "Electric Railway Not Creation of One Man," *The New York Times,* Sept. 23, 1928.

13. Letter to Frank J. Sprague from Congressman Randolph Perkins, Mar. 31, 1928, Frank Julian Sprague Papers, Mss. & Archives Section, N.Y.P.L.

14. Margaret Cousins, *The Story of Thomas*

Alva Edison (New York: Landmark Books, Random House, 1965, renewed 1993), 128–130.

15. *Frank J. Sprague and the Edison Myth.*

16. Edison, *The Sun,* Sept. 4, 1919.

17. Frank J. Sprague, "Mr. Edison's Greatest Work," *Electrical Engineering* 50, no. 12 (Dec. 1931): 977–978.

18. Apropos the New York Subway and other electric railway problems, and a few facts RE Sprague in connection therewith. Undated. Frank Julian Sprague Papers, Seashore Trolley Museum.

19. "Re; Sprague inventions and developments." Frank Julian Sprague Papers, Mss. & Archives Section, N.Y.P.L.

20. Apropos the New York Subway and other electric railway problems, and a few facts RE Sprague in connection therewith. Undated. Frank Julian Sprague Papers, Seashore Trolley Museum.

21. "Father of the Trolley Car," *New York Herald Tribune,* Oct. 26, 1934.

SELECTED BIBLIOGRAPHY

BOOKS

A Century of Progress: The General Electric Story, 1876–1978. Schenectady, N.Y.: Hall of History Foundation, 1981.

A.E.R.A., Golden Anniversary, Guest of Honor, Frank Julian Sprague. Atlantic City, N. J., Sept. 28, 1931.

Aldrich, Mark. *Death Rode the Rails: American Railroad Accidents and Safety, 1882-1965.* Baltimore, Md.: Johns Hopkins University Press, 2006.

Amended General Description, Sprague Auxiliary Train Control, Universal Selective Type of Equipment. Bulletin No. 3, March 15, 1923. New York: Sprague Safety Control & Signal Corp.

Automatic Train Control. Scranton, Pa.: International Textbook Co., 1930.

Benjamin, Park. *The United States Naval Academy.* New York: G. P. Putnam's Sons, 1900.

Burlingame, Roger. *Engineers of Democracy: Inventions and Society in Mature America.* New York: Charles Scribner's Sons, 1940.

Carlson, Stephen P. and Fred W. Schneider III. *PCC: The Car that Fought Back.* Interurbans Special 64. Glendale, Calif.: Interurban Press, 1980.

Clark, Ronald W. *Edison: The Man Who Made the Future.* New York: G. P. Putnam's Sons, 1977.

Condit, Carl W. *American Building Art: The Nineteenth Century.* New York: Oxford University Press, 1960.

Cousins, Margaret. *The Story of Thomas Alva Edison.* New York: Landmark Books, Random House, 1965, renewed 1993.

De Borchgrave, Alexandra Villard and John Cullen. *Villard: The Life and Times of an American Titan.* New York: Doubleday, 2001.

Directions for Setting Up and Running Sprague Electric Motors. New York: Sprague Electric Railway & Motor Co., 1889.

Emmet, William Le Roy. *The Autobiography of an Engineer.* Albany, N.Y.: Fort Orange Press, 1931.

Ford, Robert S. *Red Trains in the East Bay.* Special no. 65. Glendale, Calif.: Interurbans ,1977.

Frank J. Sprague: Seventy-fifth Anniversary. New York: 1932.

Goodwin, Jason. *Otis: Giving Rise to the Modern City.* Chicago: Ivan R. Dee, Publisher, 2001.

Gray, Lee E. *From Ascending Rooms to Express Elevators: A History of the Passenger Elevator in the 19th Century.* Mobile, Ala.: Elevator World, Inc., 2002.

Hammond, John Winthrop. *Men and Volts.* New York: J. B. Lippincott Co., 1941.

Hood, Clifton. *722 Miles: The Building of the Subways and How They Transformed New York.* New York: Simon & Schuster, 1993.

Israel, Paul. *Edison: A Life of Invention.* New York: John Wiley & Sons, 1998.

John Fritz Medal. Biography of Frank Julian Sprague. New York: October 1935.

Klein, Maury. *The Life and Legend of E. H. Harriman.* Chapel Hill: University of North Carolina Press, 2000.

Long, W. Earl. *Richmond 1888: Dawn of the Street Railway Era.* Richmond, Va.: Lady Liberty Press, 1988.

Martin, Thomas Commerford, and Joseph Wetzler. *The Electric Motor and Its Applications,* 2nd ed. New York: W. J. Johnston, Publisher, 1888.

McCabe, James D. *New York by Sunlight and Gaslight.* New York: Union Publishing House, 1886.

Mechanical and Electrical Data Regarding the Sprague Motor and the Transmission of Power. New York: Sprague Electric Railway & Motor Co., ca 1886.

Middleton, William D. *Grand Central: The World's Greatest Railroad Terminal.* San Marino, Calif.: Golden West Books, 1977.

———. *Metropolitan Railways: Rapid Transit in America.* Bloomington: Indiana University Press, 2003.

———. "Technology and Operating Practice in the Twentieth Century." In *Encyclopedia of North American Railroads.* Middleton, William D., George M. Smerk, and Roberta L. Diehl, eds. Bloomington: Indiana University Press, 2007: 53–64.

———. *The Time of the Trolley.* Milwaukee, Wisc.: Kalmbach Publishing Co., 1967.

———. *When the Steam Railroads Electrified.* 2nd ed. Bloomington: Indiana University Press, 2001.

Miller, Nathan. *The U.S. Navy: A History,* 3rd ed. Annapolis, Md.: Naval Institute Press, 1977, 1990, 1997.

———. *The U.S. Navy: An Illustrated History.* New York and Annapolis, Md.: American Heritage Publishing Co. and the United States Naval Institute Press, 1977.

Moffat, Bruce G. *The "L": The Development of Chicago's Rapid Transit System, 1888–1932.* CERA Bulletin 131. Chicago: Central Electric Railfans' Assn., 1995.

Passer, Harold C. *The Electrical Manufacturers: 1875–1900: A Study in Competition, Entrepreneurship, Technical Change, and Economic Growth.* Cambridge, Mass.: Harvard University Press, 1953.

———. "Frank Julian Sprague: Father of Electric Traction, 1857–1934." In *Men in Business.* William Miller, ed. Cambridge, Mass.: Harvard University Press, 1952.

Scott, Lloyd N. *Naval Consulting Board of the United States.* Washington, D.C.: Government Printing Office, 1920.

Sprague, Frank Julian. *Early Steps in the Development of Electric Traction.* Reprinted from AERA, February, 1932.

———. *Report on the Exhibits at the Crystal Palace Electrical Exhibition, 1882.* General Information Series, No. II: Information From Abroad. Navy Department, Bureau of Navigation, Office of Naval Intelligence, 1884. Washington, D.C.: Government Printing Office, 1883.

———. *The Growth of Electric Railways: A Historical Review of the Physical Development of One of the Nation's Greatest Industries.* New York: American Electric Railway Assn., 1916.

Sprague, Harriet. *Frank J. Sprague and the Edison Myth.* New York: William-Frederick Press, 1947.

Stevens, John R. *Pioneers of Electric Railroading: Their Story in Words and Pictures.* Headlights Volumes 51 and 52, 1989–1990. New York: Electric Railroaders' Assn., 1991.

Sweetman, Jack. *The U.S. Naval Academy: An Illustrated History.* Annapolis, Md.: Naval Institute Press, 1979.

The New York Central Electrification. General Electric Company, GEA-902. Schenectady, N.Y.: January, 1929.

The Sprague Electric Company's Multiple Unit System of Train Operation. Pamphlet No. 501, October, 1901. New York: Sprague Electric Co., 1901.

The Sprague System of Electric Railways as Exemplified in the Richmond Road. New York: Sprague Electric Railway & Motor Co., ca 1889.

Warren, Mame, and Marion E. Warren. *Everybody Works But John Paul Jones: A Portrait of the U.S. Naval Academy, 1845–1915.* Annapolis, Md.: Naval Institute Press, 1981.

White, John H. *Horse Cars: City Transit before the Electrical Age.* Oxford, Ohio: Miami University Libraries, 2006.

PERIODICALS & NEWSPAPERS

General

"Absorption of Sprague Electric by General Electric." *Electrical World and Engineer* 29, no. 22 (May 31, 1902): 971–972, 978.

"Engineering Professions Honor 'The Father of Electric Traction'." *A.E.R.A.* 23 (August 1932): 1196–1198.

Hedley, Frank. "An Engineer's Contribution to Transportation." *A.E.R.A.* 23 (August 1932): 1206–1210.

Holden, James W. "The Father of Electric Traction." *Railroad Magazine* 28, no. 2 (Aug. 1940): 6–24.

Jackson, Dr. Dugald C. "Frank Julian Sprague, 1857–1934," (reprint). *The Scientific Monthly* 57 (Nov. 1943): 431–441.

"John Fritz Medal for 1935 Presented to a Son of the Late Frank J. Sprague." *Electrical Engineering* 54, no. 2 (Feb. 1935): 242–243, 245.

"Memoir of Frank Julian Sprague." *Transactions of the American Society of Civil Engineers* 100 (1935): 1736–1741.

Middleton, William D. "A Century of Cable Cars." *American Heritage* 36, no. 3 (Apr./May 1985): 90–101.

"Naval Consulting Board Personnel." *Engineering Magazine* 50, no. 2 (Nov. 1915): 199–221.

Potter, Robert D. "Father of Rapid Transit." *New York Herald Tribune* (July 24, 1932).

Sprague, Frank Julian. "Coming Development of Electric Railways" (4 parts). *The Electrical Engineer* 13 and 14, nos. 215, 216, 217, and 218 (June 15, 22, 29, and July 6, 1892): 611–612, 629–630, 661–662, 712.

———. "Digging in 'The Mines of the Motors'." *Electrical Engineering* 53, no. 5 (May 1934): 695–707.

———. "Electric Traction in Space of Three Dimensions," (reprint). *The Journal of the Maryland Academy of Science* 2, nos. 3 and 4; Vol. 3, nos. 1, 2, and 3 (December 1931, July 1932,): 161–194, 1–36.

———. "Fifty-eight Years on the Firing Line." *A.E.R.A.* 23 (August 1932): 1199–1205.

———. "Mr. Sprague Discusses Early Steps in the Development of Electric Traction." *A.E.R.A.* (American Electric Railway Assn.) 23 (February 1932): 848–853.

———. "Some Application of Electric Transmission." *Scientific American Supplement,* No. 700 (Aug. 3, 1889): 11328–11330.

———. "Some Personal Experiences." *Street Railway Journal* 24, no. 15 (Oct. 8, 1904): 566–575.

———. "The Electric Railway, First Paper, A Résumé of the Early Experiments, and Second Paper, Later Experiments and Present State of the Art." *The Century Magazine* 70, nos. 3 and 4 (July and Aug. 1905): 434–451, 512–517.

———. "The Transmission of Power by Electricity." *Journal of the Franklin Institute* 127, nos. 3 and 4 (March, April 1889): 161–264.

"Street Railway Engineers—XXXIII. Frank J. Sprague." *Electric Railway Gazette* (Jan. 11, 1896): 31.

"The New Sprague Company." *The Electrical Engineer* 24, no. 493 (Oct. 14, 1897): 355.

"The Sprague System of Electric Motors." *The Electrical World* 8, nos. 16, 17, and 18 (Oct. 16, 23, and 30, 1886): 189–190, 199–200, 212.

Electric Railways

Guy, George H. "A Memorable Anniversary." *The Electrical World* 31, no. 7 (Feb. 12, 1898): 221.

Martin, T. C. "Electric Street Cars." *Transactions of the American Institute of Electrical Engineers* 4 (1887): 26–57.

Sprague, Frank Julian. "Application of Electricity to Propulsion of Elevated Railroads." *The Electrical World* 7, nos. 3, 4, 10, and 11 (Jan. 16 and 23, March 6 and 13, 1886): 27–28, 36, 106–107, 118–119.

———. "Birth of the Electric Railway." *Transit Journal* 78, no. 10 (Sept. 15, 1934): 317–324.

———. "Considerations Which Should Govern the Selection of a Rapid Transit System." *Transactions of the American Institute of Electrical Engineers* 8 (1891): 331–350.

———. "Lessons of the Richmond Electric Railway." *The Engineering Magazine* 7, no. 6 (September 1894): 787–805.

"The Sprague Electric Railway System." *The Electrical World* 8, no. 13 (Sept. 25, 1886): 151–155.

Elevators

"Electric Control Permits Two Elevators in Same Shaft." *Electrical World* 97 (Feb. 14, 1931): 312–316.

"Electricity on the Central London Underground." *The Electrical Engineer* 25, no. 515 (Mar. 17, 1898): 299–301.

Fortune, James W. "Mega High-Rise Elevators." *Elevator World* 43, no. 7 (July 1995): 63–69.

Gray, Dr. Lee. "Two Cars in One Shaft." *Elevator World* 54, no. 5 (May 2006): 2–7.

"Higher Car Speeds Proposed for Elevators." *Power* (Jan. 6, 1931).

"Passenger Lifts of the Central London Railway." *Engineering* (Mar. 3 and 10, 1899): 273–276, 305–307.

Paumgarten, Nick. "Up and Then Down." *The New Yorker* (April 21, 2008): 106–115.

Sakita, Masami. "Elevator System with Multiple Cars in One Hoistway." *Elevator World* 54, no. 6 (June 2001): 80–91.

Sprague, Frank Julian, and Charles R. Pratt. "The Sprague-Pratt Electric Elevator for High Duty Service." *The Electrical Engineer* 14, no. 235 (Nov. 2, 1892): 417–427.

———. "Electric Elevators, with Detailed Description of Special Types." *Transactions of the American Institute of Electrical Engineers* 13 (1897): 3–67.

"The Postal Telegraph Cable Company's New Building." *The Electrical World* 23, no. 16 (April 21, 1894): 525–541.

"XI.—The Sprague-Pratt Electric Elevators." *The Electrical Engineer* 17, no. 311 (April 18, 1894): 351–357.

Multiple Unit Control

"A French Sprague Company." *Street Railway Journal* 16, no. 42 (Oct. 20, 1900): 1032–1033.

Cravath, J. R. "New Electrical Equipment of the South Side Elevated Railroad, Chicago." *The Electrical Engineer* 25, no. 522 (May 5, 1898): 475–481.

"Electrical Equipment of the Brooklyn Elevated Railroad." *The Electrical Engineer* 25, no. 513 (Mar. 3, 1898): 227–229.

Hill, George H. "Some Notes on the History and Development of the Multiple-Unit System of Train Operation." *Street Railway Journal* 17, no. 18 (May 4, 1901): 551–554.

Sprague, Frank Julian. "How Multiple-Unit Control Originated." *Transit Journal* 78, no. 10 (Sept. 15, 1934): 333–334.

———. "The Multiple Unit System of Electric Railways." *Transactions of the American Institute of Electrical Engineers* 16 (1900): 193–250.

"Sprague Multiple Unit Control and the Adoption of Electricity on the Brooklyn Elevated System." *The Electrical Engineer* 25, no. 515 (Mar. 17, 1898): 301–302.

"The New Sprague Multiple Unit System of Car Traction." *The Electrical Engineer* 24, no. 482 (July 29, 1897): 93–94.

"The Sprague Multiple Unit Traction System." *The Electrical Engineer* 25, no. 483 (Aug. 5, 1897): 114.

Trunk Line Electrification

"A Large Electric Locomotive." *The Electrical World* 25 (April 6, 1895): 425.

"Comparative Tests of Consolidation and Mallet Locomotives on Southern Pacific." *Railway Age Gazette* 48, no. 2 (Jan. 14, 1910): 91–94.

"Electrification on the Southern Pacific." *The Railroad Gazette* 43, no. 10 (Sept. 6, 1907): 249–250.

"Engineering Board to Report on Electrification for Southern Pacific." *The Railway Age* 44, no. 10 (Sept. 6, 1907): 329.

General News Section, on Southern Pacific Electrification. *Railroad Age Gazette* 47, no. 3 (July 16, 1909): 113.

General News Section, on Southern Pacific Electrification. *Railroad Age Gazette* 47, no. 4 (July 23, 1909): 159.

General News Section, on Southern Pacific Electrification. *Railway Age Gazette* 48, no. 2 (Jan. 14, 1910): 101.

"Harriman and the Southern Pacific." *Street Railway Journal* 29, no. 4 (Jan. 26, 1907): 165.

"Southern Pacific Company." *Street Railway Journal* 30, no. 24 (Dec. 14, 1907): 1154.

"Southern Pacific Improvements." *Street Railway Journal* 30, no. 19 (Nov. 9, 1907): 972.

"Southern Pacific to Consider Electrification." *Street Railway Journal* 30, no. 9 (Aug. 31, 1907): 325.

Sprague, Frank Julian. "On the Substitution of the Electric Motor for the Steam Locomotive." *Street Railway Journal* 29, no. 6 (Feb. 9, 1907): 249–250.

———. "High Voltage Direct Current Traction." *Railway Age Gazette* 49, no. 16 (Oct. 14, 1910): 688.

———. "Some Facts and Problems Bearing on Electric Trunk Line Operation." *Transactions of the American Institute of Electrical Engineers* 26, Part 1 (1908): 681–812.

———. "Trunk Line Electrification." *Street Railway Journal* 29, no. 21 (May 25, 1907): 907–918.

"The 1,000 H.P. Sprague Electric Locomotive." *The Electrical Engineer* 16, no. 285 (Oct. 18, 1892): 339–341, 346.

Wilgus, William J. "The Electrification of the Suburban Zone of the New York Central and Hudson River Railroad in the Vicinity of New York City." *Transactions of the American Society of Civil Engineers* 61 (December 1908): 73–155.

Automatic Train Control

"Automatic Train Stops in New York." *Railway Age* 72, no. 6 (Feb. 11, 1922): 305.

"Block Signal Mileage, 1928." *Railway Age* 85, no. 15 (Oct. 13, 1928): 706.

Ellis, G. E. "The Status of Train Control." *Railway Age* 83, no. 2 (July 9, 1927): 69–70.

Ellis, G. E., et al. "Automatic Train Control." *Proceedings, The Journal of the Pacific Railway Club* 11, no. 4 (July 1927): 3–52.

"Four Objections to Installing Automatic Stops." *Railway Age* 72, no. 12 (Mar. 25, 1922): 783–784.

"Great Northern Train Control." *Railway Signaling* 18, no. 1 (Jan. 1925): 6–11.

ICC report on track signaling mileage, Jan. 1, 1908. *The Railroad Gazette* 44, no. 16 (Apr. 17, 1908): 543–548.

Interstate Commerce Commission Reports, 258–281, Order No. 13413, dated June 13, 1922.

Interstate Commerce Commission Reports, Order No. 13413, dated June 13, 1922, In the Matter of Automatic Train-Control Devices.

Latimer, J. B., "Train Operation and Automatic Train Control." *Railway Age* 70, no. 16 (Apr. 22, 1921): 977–979.

"New York Commission on Automatic Train Control." *Railway Age* 70, no. 9 (Mar. 4, 1921): 497–498.

"Rapid Progress in Automatic Train Control in 1926." *Railway Age* 82, no. 1 (Jan. 1, 1927): 130–133.

"Report on Sprague Automatic Train Control." *Railway Age* 75, no. 9 (Sept. 1, 1923): 399–400.

"Signals and Automatic Train Stops in the Hudson & Manhattan Tunnel." *The Railroad Gazette* 44, no. 10 (Mar. 10, 1908): 317.

Sprague, Frank Julian. "Automatic Train Control." *Journal of The Franklin Institute* 194, no. 2 (August 1922): 133–163.

———. "The Need for Automatic Train Control." *Railway Age* 68, no. 6 (Feb. 6, 1920): 401–402.

Sprague Auxiliary Train Control, Bulletin No. 3 (Mar. 15, 1923).

"The Block Signal and Train-Control Board." *The Railroad Gazette* 44, no. 2 (Jan. 10, 1908): 52–53.

"The New Train Control Order." *Railway Age* 76, no. 5 (Feb. 2, 1924): 314.

"The Sprague System of Automatic Train Control." *Railway Review* (May 27, 1922): 747–756.

"What the Roads Have Done with Train Control." *Railway Age* 76, no. 12 (Mar. 12, 1924): 637–639.

INDEX

Books in the *Railroads Past & Present* Series

Landmarks on the Iron Railroad: Two Centuries of North American Railroad Engineering by William D. Middleton

South Shore: The Last Interurban (revised second edition) by William D. Middleton

Katy Northwest: The Story of a Branch Line by Don L. Hofsommer

"Yet there isn't a train I wouldn't take": Railroad Journeys by William D. Middleton

The Pennsylvania Railroad in Indiana by William J. Watt

In the Traces: Railroad Paintings of Ted Rose by Ted Rose

A Sampling of Penn Central: Southern Region on Display by Jerry Taylor

The Lake Shore Electric Railway by Herbert H. Harwood, Jr., and Robert S. Korach

The Pennsylvania Railroad at Bay: William Riley McKeen and the Terre Haute and Indianapolis Railroad by Richard T. Wallis

The Bridge at Quebec by William D. Middleton

History of the J. G. Brill Company by Debra Brill

Uncle Sam's Locomotives: The USRA and the Nation's Railroads by Eugene L. Huddleston

Metropolitan Railways: Rapid Transit in America by William D. Middleton

Perfecting the American Steam Locomotive by J. Parker Lamb

From Small Town to Downtown: A History of the Jewett Car Company, 1893–1919 by Lawrence A. Brough and James H. Graebner

Limiteds, Locals, and Expresses in Indiana by Craig Sanders

Amtrak in the Heartland by Craig Sanders

When the Steam Railroads Electrified by William D. Middleton

The GrandLuxe Express: Traveling in High Style by Karl Zimmermann

Still Standing: A Century of Urban Train Station Design by Christopher Brown

The Indiana Rail Road Company by Christopher Rund

Evolution of the American Diesel Locomotive by J. Parker Lamb

The Men Who Loved Trains: The Story of Men Who Battled Greed to Save an Ailing Industry by Rush Loving

The Train of Tomorrow by Ric Morgan

Built to Move Millions: Streetcar Building in Ohio by Craig R. Semsel

The CSX Clinchfield Route in the 21st Century by Jerry Taylor and Ray Poteat

Visionary Railroader: Jervis Langdon Jr. and the Transportation Revolution by H. Roger Grant

The New York, Westchester & Boston Railway: J. P. Morgan's Magnificent Mistake by Herbert H. Harwood, Jr.

313

Iron Rails in the Garden State: Tales of New Jersey Railroading by Anthony J. Bianculli

Visionary Railroader: Jervis Langdon Jr. and the Transportation Revolution by H. Roger Grant

The Duluth, South Shore & Atlantic Railway by John Gaertner

Iowa's Railroads: An Album by H. Roger Grant and Don L. Hofsommer

WILLIAM D. MIDDLETON graduated with a degree in civil engineering from Rensselaer Polytechnic Institute and is a professional engineer. His career includes work as a structural engineer and bridge designer, a long-time officer in the U.S. Navy's Civil Engineer Corps, and an extensive assignment as chief facilities officer at the University of Virginia. His publications include more than 700 magazine articles and more than 20 books. He is editor of (with George M. Smerk and Roberta L. Diehl) and a contributor to the *Encyclopedia of North American Railroads* (Indiana University Press, 2007).

WILLIAM D. MIDDLETON III is an archaeologist who has studied at the University of California, San Diego, San Francisco State University, and the University of Wisconsin, where he earned his Ph.D. in 1998. His professional work has included teaching and conducting research at the University of Wisconsin, Fond du Lac; the Field Museum of Natural History; DePaul University; and Rochester Institute of Technology. His published work includes numerous professional papers, and he is a contributor to the *Encyclopedia of North American Railroads* (Indiana University Press, 2007).

This book was designed by Jamison Cockerham and set in type by Tony Brewer at Indiana University Press and printed by Sheridan Books. The sponsoring editor was Linda Oblack and the project editor was June Silay.

The text face is Arno Pro, designed by Robert Slimbach, and the display face is Caecilia LT Std, designed by Peter Matthias Noordzij, both issued by Adobe Systems.